COSMETIC MICROBIOLOGY

A Practical Handbook

edited by

Daniel K. Brannan

CRC Press
Boca Raton New York

Publisher, Life Sciences: Robert B. Stern
Editorial Assistant: Carol Messing
Project Editor: Carrie L. Unger
Cover design: Dawn Boyd
PrePress: Carlos Esser

Library of Congress Cataloging-in-Publication Data

Cosmetic microbiology : a practical handbook / edited by Daniel K. Brannan
p. cm.
Includes bibliographical references and index.
ISBN 0-8493-3713-5 (alk. paper)
1.Cosmetics--Microbiology. I. Brannan, Daniel K.
QR53.5.C76C67 1997
668′.55--dc21

96-40142
CIP

Direct all inquiries to CRC Press LLC, 2000 Corporate Blvd., N.W., Boca Raton, Florida 33431.

International Standard Book Number 0-8493-3713-5
Library of Congress Card Number 96-40142
Printed in the United States of America 1 2 3 4 5 6 7 8 9 0
Printed on acid-free paper

PREFACE

My vision in editing this book was to integrate all practical aspects of cosmetic microbiology into an understandable and readable text. Many of the people who developed cosmetic microbiology in the 1950s and 1960s are now retired from the industry. Much of the knowledge they had on the subject took form in the "oral tradition" passed on to their successors, who typically served lengthy "apprenticeships" before taking over.

This book is a culmination of that quest. The reader should find it technical enough to educate new microbiologists coming into the field, thus filling the void of the traditional oral mentors who are rapidly disappearing. However, it should still be readable and simple enough so that the process engineers, plant managers, and workers can learn the essential items they need to create safe products. It is an ideal text for corporate, technical training workshops.

I believe this book fills a niche that is currently void. There are books on food microbiology (Troller's *Sanitation in Food Processing* published by Academic Press, 1983), and there is an exceptional book on pharmaceutical microbiology (Hugo and Russell's *Pharmaceutical Microbiology, Third Edition* published by Blackwell Scientific, 1983). Certainly there are also some books on cosmetic microbiology and cosmetic preservation, but these are either monographs promoting a rather biased set of ideas or they are little more than preservative suppliers' advertising dogma rather than a discussion that is scientific and technically sound.

I hope this book will evolve into a classic analogous to Hugo and Russell's *Pharmaceutical Microbiology*, only for the cosmetic microbiologist. In order for that to happen, we will need additional contributors to provide additional chapters in future editions. It is to those readers, who may look over this book and realize that they could have done a better job, that I appeal. Go ahead, write it down, and send it to me. It will likely find its way into the second edition!

THE EDITOR

Daniel K. Brannan, Ph.D., is Chair and Professor of Biology at Abilene Christian University (ACU) in Abilene, Texas. Since leaving a career in the industry, he has served as a consultant for dozens of cosmetic and drug companies both nationally and internationally.

Dr. Brannan graduated in 1975 from the University of New Mexico, Albuquerque, New Mexico, with a Bachelor of Science degree in biology with distinction and with the honor of being elected to Phi Beta Kappa. He then obtained his Master's of Science degree in microbiology in 1977 from Ohio State University, Columbus, Ohio. In 1981, he received his doctorate degree in biology from the University of New Mexico. He worked for the Procter & Gamble Company from 1981 to 1988 and has taught at ACU since 1988.

Dr. Brannan is a member of the American Society for Microbiology (ASM), American Association for the Advancement of Science (AAAS), Society for Industrial Microbiology, and American Association of University Professors. He served on the Cosmetic, Toiletries and Fragrance Association Methodology and Microbiology Committees from 1983 to 1987. He also served on the U.S. Pharmacopeial Microbiology Advisory Panel on Microbiological Control from 1990 to 1995. He is a reviewer for the *Journal of the Society of Cosmetic Chemists*, as well as being a member of the ASM Board of Education and Training Committee for Continuing Education. Finally, he is the science advisor for the Regional HIV Prevention Coalition of the Texas Department of Health.

Dr. Brannan has been the recipient of nearly $300,000 in research grants from the Environmental Protection Agency, the Department of Energy, the Research Council, and the Texas Higher Education Coordinating board. He has presented at more than 35 invited lectures and workshops at international and national meetings. He has also published numerous papers. His current research interests include biofilms and resistance mechanisms of bacteria to biocides and preservatives.

CONTRIBUTORS

William A. Apel, Ph.D.
Idaho National Engineering Laboratories
Idaho Falls, Idaho

Daniel K. Brannan, Ph.D.
Department of Biology
Abilene Christian University
Abilene, Texas

Mary K. Bruch, B.S.
Myrtle Beach, South Carolina

Janet C. Curry, B.S.
Microbiological Quality Control, Inc.
Woodrige, New Jersey

Philip A. Geis, Ph.D.
The Procter & Gamble Company
Cincinnati, Ohio

Richard T. Hennessey, M.S.
The Procter & Gamble Company
Cincinnati, Ohio

Mary Ann Magee, B.S.
Bausch & Lomb Healthcare
Rochester, New York

Richard Mulhall, B.S.
Alberto-Culver
Melrose Park, Illinois

Edward Schmidt, M.P.H.
Alberto-Culver
Melrose Park, Illinois

Thomas E. Sox, Ph.D.
Merk Pharmaceuticals
Ambler, Pennsylvania

Scott V. W. Sutton, Ph.D.
Optics Worldwide
Alcon, Inc.
Ft. Worth, Texas

TABLE OF CONTENTS

COSMETIC MICROBIOLOGY

A Practical Handbook

SECTION ONE: GENERAL CONCERNS AND BASIC SANITATION

1 HISTORY OF COSMETIC MICROBIOLOGY

JANET C. CURRY

CONTENTS

INTRODUCTION

A fascinating study is the history of cosmetics and microbiology and how they merged into a new field of microbiology. People have a basic need to change their appearance. The vastness of today's cosmetics and toiletries industry clearly indicates this widespread and basic need for cosmetics. Perhaps this need arises because cosmetics allow us to make ourselves unique for rituals or societal reasons. This phenomenon dates to our beginnings, suggesting a behavioral characteristic that may even border on instinct. We apparently have a primal need for cosmetics to provide for our well-being. Cosmetics are cures for the disease of being someone whom we do not want to be.

This chapter provides a panoramic view of the history of cosmetics and microbiology as a scientific discipline. Within it, we also explore the need to combine the two disciplines. The intent is to provide insight into this unique industry to which we belong.

0-8493-3713-5/97/$0.00+$.50

THE FIRST COSMETICS

Archeological evidence suggests that we have been decorating our bodies for nearly 500,000 years. Archaeologists discovered red clay sticks for coloring the body with ocher. Based on current scientific ideas, this would have occurred while our earliest ancestor, *Homo erectus*, occupied the African savannah between 300,000 and 1.5 million years ago. *Homo erectus* was a hominid that ventured from the forest onto the African savannah.

According to anthropologists this hominid shared many characteristics with us. Current scientific thought is that *Homo erectus* had twice the brain size of other hominids, a body adapted to upright posture, less hair, and sweat glands. *Homo erectus* also developed a workable social life, giving them an edge over other hominids. The men were hunters who first used stones and clubs as tools and weapons. Later on, they used chipped stones and spears. Their women were gatherers who took care of the children while sharing their food at day's end. These hominids may have spent the next 800,000 years enjoying their success.

They moved around as family units and joined with other families if needed. The adventurous families moved from the tropics to the colder climates of West Asia, Europe, India, and China during several ice ages. Clothing and shelter became necessary. Man then captured fire, worshipped it, and kept it alive for generations. Fire was very important since it provided fuel for warmth, cooking, and hardening of wooden spear tips.

The life of *Homo erectus* was tougher. Nature was more frightening. Magic, wizardry, the supernatural, and various ritualistic ceremonies were popular male pastimes. Those mysterious red clay sticks and charcoal may have been for body painting. This painting could have been a part of secret male rites performed deep within caves. *Homo erectus* then disappeared around 300,000 years ago.

Red clay remained popular throughout succeeding Stone Ages. The Neanderthals, a variety of *Homo sapiens*, lived 30,000 to 125,000 years ago. They painted their dead with red ocher before burial. This practice may have been done to bring the blush of life back to the body. Then, Cro-Magnons appeared around 40,000 years ago, with a skull shape more conducive to brain development.

Cro-Magnons are modern humans while Neanderthals had heavier brow ridges and were more primitive. Cro-Magnon, like modern man, had a round head and high brow. He moved quickly throughout the world, and replaced the Neanderthals. Unopposed, they spent the next 40,000 years evolving into the right social structure of modern man.

They left a rich legacy of cave art, animal carvings, baked clay sculptures, and those ever-present red clay sticks. They lived together in clans or tribes. Cro-Magnons trained and bred animals for pets as well as beasts of burden. They spun natural fibers into cord for baskets, bows and arrows, fishing lines, and animal snares. Once the last Ice Age ended nearly 10,000 years ago, they no longer moved from place to place in search of food. This ended the Paleolithic Age and the nomadic family life style, and it began the construction of villages and towns with the need for specialized occupations. The end of the nomadic culture also began complex communal cultures.

The start of farming marked the beginning of what we call the New Stone (or Neolithic) Age. This age occurred at different times in different parts of the world. It started in Southwest Asia nearly 10,000 years ago. In Mexico and South America, it began as late as 1500 B.C. The discovery of copper and tin began the Bronze Age. This age occurred at different times in different places; the earliest age in Southwest Asia and Egypt was around 3500 B.C.

The need for highly specialized occupations in these emerging cultures may have led to the development of professional classes, upper and lower social classes. It may even have led to territorial wars. Civilization's arrival was not very civil. Experts define civilization as the time when man developed a written language. This first language occurred about 3200 B.C. in Sumer, an area in present-day Iraq between the Tigris and Euphrates Rivers. The people there developed cuneiform symbols carved into clay tablets. This was near the same area from which the Jewish creation story of Adam and Eve in Eden as told. According to the story, once their eyes were opened and they knew what they looked like, they began adorning themselves. This story emphasizes our basic need to alter our appearance.

First settled around 4500 B.C., Sumer became a land of cities within about 1000 years. The Sumerians were hardworking, cooperative people who refined farm practices to include large networks of irrigation canals. They had more food than they needed. As a result, they developed a brisk trading network with Indian and Egyptian merchants that created a need for a written language.

By 2700 B.C., the ruling class of Egyptians in the lush Nile Valley was enjoying luxurious and leisurely lives at the expense of slaves. Their cities were more elaborate. They built great pyramids to house the royal dead. Upper and lower classes were well-defined. The wealthy wore sheer linens instead of sheepskins. Their language was artistically beautiful as hieroglyphics. They wrote elegantly on papyrus instead of clay tablets. And they remain famous, even today, for their cosmetics, hair dyes, wigs, and other adornments.

During this time, man refined cosmetics into luxuries. The lower classes had no leisure time for such frivolities. We also could view this time as the emergence of the basic need to separate oneself from the lower classes by changing one's appearance.

Therefore, *Homo erectus* was apparently the first to use red clay sticks for ritualistic body painting nearly 500,000 years ago. But the more elegant kohl, henna, pomades, and hair dyes were used by wealthy Egyptian women 5000 years ago. Painting our skin is a part of human nature. If this activity is a human characteristic, and if clay was used 500,000 years ago as our first cosmetic, then cosmetics have been with us throughout the development of culture.

THE FIRST BACTERIA

Some scientists think that bacteria were the first living things on Earth. According to current scientific thinking, they either came from reducing organic chemicals in watery soups, or were bound to crystalline clay surfaces about 3.5 billion years ago. From that point, they evolved into different forms and adapted to whatever environment came their way. They saw lots of changes. Earth cooled, land masses and oceans formed, and algae produced oxygen. Bacteria continued adapting as they formed consortia of organisms that acted as if they were a single entity. Over time, these consortia, according to the endosymbiont theory, developed into plants, animals, and eventually humans.

Bacteria still control the earth as we know it. They are in and on every environmental niche where water is liquid on Earth. They even occupy us. Some are disease-producing, but most are not. Some are even needed for our existence. If all bacteria were suddenly to die, all other living creatures would also die. However, if all the plants and animals were to die, the bacteria would still survive and adapt to their new environment to start a new cycle. This is a humbling thought. It is well worth remembering as we approach the short 300-year history of microbiology, and the even shorter 70-year history of cosmetic microbiology.

MICROBIOLOGY AS A SCIENTIFIC DISCIPLINE

In the 13th century, the English friar Roger Bacon claimed that "invisible creatures" caused disease. It was not until the late 1600s that a wealthy linen merchant, Anton van Leeuwenhoek, developed the microscope to reveal the existence of an entire universe of tiny living creatures. He was the first to see microorganisms. Because of his careful drawings of these microorganisms, most microbiologists consider him the Father of Microbiology.

About 100 years later (1798), a country doctor named Jenner noticed that farmers and milkmaids who contacted cows infected with cowpox were less apt to contract smallpox. He reasoned that exposure to cowpox would strengthen the body's defenses. He then made history by rubbing cowpox exudate into the skin of an 8-year-old boy proving that immunization worked. As a result, he laid the foundations of modern-day immunology.

Another 82 years passed before Pasteur improved the vaccination process (1880). He used attenuated cultures in his work with chicken cholera. He named the process vaccination based on the word, vaca, (Latin for cow) to honor Jenner's use of cowpox. Soon after this, the American Theobald Smith, improved the process. He used heat-killed cultures, and this remains the basis of today's vaccine prophylaxis.

Microbiology was primarily a curiosity for 200 years after van Leeuwenhoek. Then, after 1850, microbiology exploded into a scientific discipline. This surge in interest was mainly due to the spontaneous generation controversy and Koch's postulates for infectious disease. In 20 years, between 1880 and 1900—the Golden Era of microbiology—scientists discovered causative organisms of just about every important disease. They isolated, described, and controlled the major bacterial pathogens and even a few viruses. Thus began the taming of killers such as anthrax, diphtheria, tetanus, typhoid fever, yellow fever, rabies, syphilis, and tuberculosis.

The major names of the period are Pasteur (France), Koch (Germany), and Lister (England), who each attracted students and scientists from all over the world. Their collective accomplishments are staggering. Within 20 years, microbiologists began understanding bacteria well enough to use and control them. American students from these European and English laboratories of the masters brought microbiology to the U.S. In 1899, the Society of American Bacteriologists (SAB) was founded. It is now known as the American Society of Microbiology (ASM), and is the largest biological scientific organization in the world.

The next 20 years were a time of maturation. Many investigators recognized the need to bring order out of the chaos generated by the activity of the previous 20 years. For example, in 1914 there was the establishment of the Digestive Ferments Company (now known as Difco). This company introduced dehydrated media, which saved not only long hours of preparation time, but provided much needed uniformity to culturing conditions.

Uniform methods and taxonomic descriptions became the order of the day. The major contributors were the American Public Health Association (APHA) and the Society of American Bacteriologists. The APHA published the first editions of *Standard Methods for Water* and *Standard Methods for Dairy Products* in the early 1900s. The first edition of *Recommended Methods for Microbiological Examination of Foods* was published in 1958. The latest editions are included in the literature cited section below; however, be aware that these are frequently revised.[1,2,3]

Other essential texts needed by today's cosmetics and pharmaceutical microbiologists are *U.S. Pharmacopeia* (XXIII, 1995) for drugs, *FDA Bacteriological Manual* (1992), and *Cosmetics, Toiletries, and Fragrances Classification (CTFA) Technical Guidelines* (1993) for cosmetics and toiletries.[4,5,6] Without uniformity of media, methods, and data expression, we could not understand or reproduce other people's work. We owe a debt of gratitude to those before us, who brought order out of chaos.

COSMETIC MICROBIOLOGY

Cosmetics and microbiology did not meet much before the 1930s. We will start with the 1930s and proceed by decades.

1930 to 1939

The cosmetics industry turned the economic disaster of the stock market crash and World War II in Europe to its advantage. Throughout history, cosmetics were expensive luxuries and the province of the wealthy. When Wall Street toppled, many cosmetic companies failed, but many others took their place. These were companies such as Merle Norman, Almay, Revlon, Clairol, Shulton, and Avon. They rapidly adjusted to the times by introducing increased advertising, new marketing methods, and new packaging. Packaging was the unique innovation. It provided the product in smaller, more affordable sizes.

Because the industry was dependent on imported raw materials unavailable due to the war, they conducted intensive research to find less expensive synthetics. This led to the development of many new and less expensive products.

Obvious proof that cosmetics had arrived was their addition to the old Food and Drug Law of 1906. Revised in 1938, it became the Food, Drug, and Cosmetic Act. Its enforcement art was the Food and Drug Administration, which was established in 1931. Most microbiological problems during this time were visible mold growth. Research on preservatives started. The parabens became well known for their antifungal activity and became the preservatives of choice.

A unique discovery during the mid-1930s was Reneé Dubos' finding of a bacterial enzyme that dissolved the polysaccharide coat of Type III pneumococcus. He isolated the coat material and used it to enrich a soil sample from a New Jersey cranberry bog. This enrichment selected one organism able to use the polysaccharide. Within 3 years, he isolated the enzyme responsible and proved it cured infected animals. The enzyme removed the polysaccharide capsule so the pneumococci were no longer protected from the body's phagocytic defense system. Unfortunately, the shift to antibiotics (needed in the war overseas) eclipsed further development of the enzyme.

What is fascinating about Dubos' discovery is that we are just now returning to his concept with the use of antimicrobial peptides. Could we use enzymes as preservatives for cosmetic products? *Pseudomonas, Klebsiella,* and *Enterobacter* also have protective polysaccharide coats. By removing the polysaccharide coat, they might die before sticking to surfaces and adapting to the product.

The 1930s also saw major microbiological advances in the medical field. The introduction of the sulfa drugs was the first advance in chemotherapy since Ehrlich's arsenical treatment of syphilis 50 years earlier. Dubos also discovered tyrothricin, a mixture of two polypeptides—tyrocidine and gramicidin—from the soil microorganism *Bacillus brevis.* It was the first combination antibiotic produced commercially and employed clinically. Gramicidin proved to be the most useful. Dubos used it to cure his first patient, Elsie the Borden cow. She had contracted mastitis at the 1939 World's Fair in New York City. The 1930s historically represent the dawn of the antibiotic era.

1940 to 1949

In the 1940s, World War II (1939 to 45) wreaked havoc on the battlefields at a cost of about 10,000,000 lives. As so often happens, however, disasters create needs. Science, medicine, and technology took giant strides forward. The search for antibiotics continued with Waksman's discovery of streptomycin in 1942. Avery's discovery of DNA as the molecular basis of heredity in 1944 provided the most far-reaching significance.

As in World War I (1914 to 18), most people perceived the cosmetic industry to be nonessential. Either we needed raw materials elsewhere or they were completely unavailable. In response to this shortage, new raw materials developed, and the industry went through another healthy period of change.

The cosmetic business continued to flourish,.primarily due to women working in factories. They had become wager earners, and their spending habits included cosmetics. Thus, there was the important economic change from perceiving cosmetics as staples rather than luxuries. The industry began to encounter some of the problems of high-volume manufacturing. Isolated instances of both spoilage and toxicity began to appear.

One reaction to this was the establishment of a Scientific Section by the CTFA (then known as the Toilet Goods Association or TGA) in 1943. It also saw M. DeNavarre found the Society of Cosmetic Chemists in 1945. DeNavarre published the first edition of "The Chemistry and Manufacture of Cosmetics" (1941). It proved to the industry that scientific discussions of formulating technicalities, chemical compatibilities, and preservatives were educational, not proprietary. This was no small accomplishment, considering the thickness of the secrecy cloak enveloping the industry.

Microbiology became more important than ever to cosmetics during the 1940s. This increased interest was due to a more scientific approach to cosmetic manufacturing. It was also due to a growing availability of test organisms. Mold spoilage remained the prime microbiological product defect to avoid. Parabens remained the preservatives of choice, but the quest increased for new ones. For the first time, preservative test procedures included bacteria.

1950 to 1959

The Korean War (1950 to 53) marred the early 1950s. But the late 1940s and early 1950s were times of prosperity, accented by the arrival of television. Television greatly affected communication and advertising of cosmetics. The 1950s were also the decade of antimicrobial overuse which continues today. Antibiotics worked miracles during the war. Physicians prescribed them for just about everything. Infections were relics of the past—or so we thought at the time. Even with the appearance of antibiotic-resistant *Staphylococcus aureus* in the hospitals, the drug industry discovered many alternatives. The antibiotic euphoria continued unabated.

The same euphoria occurred with antibacterials in toiletries. Chemists discovered halogenated bisphenolics, salicylanilides, carbanilides, and pyridinethiones, which led to a new generation of products. The function of these products was to kill the germ. There were new antibacterial skin creams, toothpastes, mouthwashes, deodorants, deodorant soaps, anti-dandruff shampoos, and surgical scrubs. These were products containing active ingredients for reducing or altering the bacterial flora on skin or in the mouth. Sometimes the antimicrobial was hidden as the preservative instead of as an active ingredient.

There are still many who consider these products to be well-preserved just because they contain antibacterial compounds. Even the annual FDA list of antibacterials used in our products does not differentiate between actives and preservatives. We list them separately because we define them by the target organisms they affect. Antimicrobials that are active ingredients primarily kill Gram-positive flora on our body surfaces. Antimicrobials that are preservatives primarily kill molds, yeasts, and Gram-negative flora in the product. But despite a product containing a biocide, one should never expect the antibacterial active to also act as a preservative or vice versa.

Between efficacy testing of the new antimicrobial products and increased attention to preservatives, microbiologists of the 1950s were hyperactive. For both types of testing, microbiologists conducted studies on the biocides to determine their spectrum of activity and the concentrations needed. Efficacy studies included various bacteriostatic, bactericidal, substantivity, and in-use panel tests. Microbiologists developed most of these tests specifically for claim support. Preservative studies included incompatibilities with product ingredients, the effects of pH, temperature, packaging, and accelerated storage conditions. There were as many different preservative challenge test protocols as there were different companies.

By the end of the decade, both industry and medicine were more aware of the growing problem of bacterial resistance. Up to this point, most control of pathogens was done with antibiotics against Gram-positive cocci. The key organisms to kill were the pneumococci (*Diplococcus pneumoniae*, later called *Streptococcus pneumoniae*), β-hemolytic streptococci (*Streptococcus pyogenes*), and staphylococci (*Staphylococcus aureus*). By the end of the decade, control of the Gram-negative bacilli also became important. Both the industry and the medical field began noticing increased problems with *Escherichia, Klebsiella, Enterobacter, Serratia*, and most importantly, *Pseudomonas*.

1960 to 1969

Many remember this period as the Vietnam era and a period of social unrest. It also produced a back-to-nature movement. The accent was on safe products with natural ingredients. Of course, no one has ever defined "natural." Environmentalists, consumerists, and women activists profoundly affected the cosmetic industry. Cosmetics were always the least regulated of all consumer products. During the 1960s, they became the target of various special interest groups. Government agencies also became more interested in them. The industry had to shed its last vestiges of secrecy. It had to open its doors voluntarily or face forced compliance with potentially unfair or impractical regulations. The FDA recognized that enforcement of any regulations would be difficult without industry cooperation. Therefore, the CTFA and FDA established a liaison and agreed on a unique program of industry self-regulation.

In 1969, a report from Sweden by Professor Kallings showed contamination in nonsterile drugs and cosmetics. He was one of the first to recognize that the industry may have microbiological safety problems. An FDA report on a survey of cosmetics from the New York area soon followed. Both reports showed contaminated cosmetics at a frequency of about 25% and a high incidence of Gram-negative bacteria. CTFA quickly established the Microbiology and Quality Assurance committees to look into the situation and to develop technical guidelines for the industry.

During this period, clinical microbiologists saw more patients with Gram-negative infections. They attributed some of these to poor quality water supplies or a laxity in sanitation practices. Other developments were even more alarming to the health-care and cosmetics industries. Contaminated disinfectants and hand cream dispensers were showing up in hospitals.

The 1960s also saw the emergence of G-11 toxicity (a halogenated bisphenol known by its nonproprietary name as hexachlorophene). This incident occurred when a product containing an exceptionally high level of G-11 (due to manufacturing error) caused the deaths of 49 infants in France. The late 1960s ended with a sense of urgency and with the industry on the defensive. We needed more effective and more responsible preservation practices to assure the microbiological and toxicological safety of the products.

1970 to 1979

The flurry of activity by the CTFA during the late 1960s quickly developed into a storm of contributions. The CTFA issued technical guidelines covering all aspects of good manufacturing and microbiological practices. The industry had taken note of the urgency to clean house. This was obvious when a CTFA national survey of almost 4000 marketplace cosmetics and toiletries over a 3-year period (1972 to 1975) showed minimal contamination. Instead of the 25% contamination incidence shown in the late 1960 surveys, the new CTFA survey proved the industry achieved a contamination rate of only 2%. Marketplace products made by reputable companies were clean. There were, and probably still are, small companies that make products without attention to microbiological control or good manufacturing practices. Both industry and the FDA continue to emphasize the need for product cleanliness and safety.

During the latter half of the 1970s, eye cosmetics came under special scrutiny. There were several cases of blindness occurring due to *Pseudomonas*-contaminated mascaras. The finding was that most products reached the consumer in good microbiological condition. But could they withstand the insult of organisms added to the product during use? This would be the main question to address over the next 15 years.

As a result, the FDA gave a contract to Georgia State University in 1975 to develop eye area challenge tests. The "double membrane technique" resulted and was further tested in another FDA contract given to the University of California at San Francisco in 1977. Again, the CTFA acted quickly. It set up an Eye Area Task Force. This group arranged a collaborative study comparing Dr. Ahern's surface-membrane preservative test with the CTFA direct-inoculation procedure. Both tests were proven to be equally satisfactory for measuring preservative activity. The industry was alert to extend its product responsibility from manufacturing all the way through to consumer use.

The pharmaceutical industry continued developing antibiotics to combat increasing Gram-negative infections. However, both industrial and clinical microbiologists were gaining new respect for microbial resistance and adaptability. Renewed attention to asepsis, hygiene, and sanitation began tempering our over-dependence on antibiotics.

1980 to 1989

The 1980s saw people facing the devastation of AIDS, the fatal virus-induced immunodeficiency disease that continues to affect people from all walks of life. On the plus side, the 1980s also saw major advances by molecular biologists. Armed with knowledge of the double helix structure of DNA since 1953, they successfully constructed recombinant DNA in the lab by 1975. With the groundwork laid, the 1980s saw explosive research efforts toward gene therapy, chromosome mapping, and agricultural trials of genetically engineered plants. Their work has had, and will have, profound effects on all future microbiological endeavors.

Directly applicable to our industry is the research for *in vitro* alternatives to animal testing. Both industry and the private sector are aggressively pursuing this avenue either with grants to universities (Procter and Gamble) or direct research (Johns Hopkins). When (or if) we can end animal testing remains to be seen. Microbiological and tissue culture methods are starting to replace traditional toxicology testing with animals. This is helping to reduce the number of animals necessary for data supporting safe human use.

The FDA also continued its campaign to develop preservative challenge methods that were predictive of consumer contamination risks. The FDA's chief criticism is that these methods do not predict if consumers can contaminate the product during use. In 1985, the FDA gave a contract to Schering-Plough and the University of North Carolina. They tried comparing preservative challenge testing methods for creams and lotions that were predictive of consumer contamination. The FDA has published these results, but they were compared to a simulated in-use regimen that is not valid as a true in-use comparison.[7] In 1987, the CTFA published the results of a survey to determine if companies tried to correlate their challenge test data with consumer use data. Nearly all companies claimed that they had correlation programs in place. None presented their data, however. There are only two challenge test methods that have published validation data proving their ability to predict the in-use potential for consumer contamination.[8,9] Both tests are modifications of the CTFA test.

1990 and Beyond

The cosmetics and toiletries industry is beginning to face the ramifications of the pending European Community (EC) open market in 1992. Multinational companies remain concerned about requirements for their products from country to country. These include uniformity of test methods, manufacturing practices, labeling requirements, and other details of the EC Commission. The 1990s continue to see a worldwide emergence of the need for industrial responsibility to protect the environment. Some key issues are clean air and water, understanding the effect of hydrocarbons and fluorocarbons on the UV protecting ozone layer, saving the diversity of flora and fauna in rain forests and other natural habitats, and properly disposing or recycling wastes. More strict regulations and laws for a world with ever-shrinking resources will be major concerns to our industry for some time to come.

Research on understanding and controlling AIDS continues. The FDA has even eased its rules concerning new drug clinical trials to restrain this devastating disease. The FDA recently granted permission to geneticists to proceed with their first gene therapy trail (with Thalassemia patients). Microbiologists are now seeing a return of diseases we thought were under control: tuberculosis, syphilis, and cholera. Antimicrobial resistance to antibiotics and even to disinfectants such as chlorine are starting to show up. These are the issues that research must resolve in the future.

CONCLUSION

These are also the issues that cosmetic microbiologists face. We now face a period of bacterial supremacy. The Gaia hypothesis of all the microorganisms on Earth behaving as one organism almost seems tenable.

Pathogens are no longer easy to define. Innocuous spoilage organisms are potential pathogens especially if present in high numbers. In 1990, the FDA established a joint program with CTFA and AOAC to develop standard preservative challenge tests that are predictive of consumer contamination. Over the next few years, we should see if this program provides any improvement over existing test methods. Regardless of the outcome, all preservative tests should aim for self-sterilization of product units, rather than simple reduction to so-called safe (10^3/g) levels. There is no safe level if survivors adapt and grow to high numbers. Bacteria have been honing their adaptation skills for billions of years. They are ready and willing to show off those skills in the industry's new milder, safer, environmentally friendly, so-called natural, products of the future. We must adjust accordingly.

We must remember that no amount of lab testing or monitoring can assure product quality in a climate of bacterial adaptation. In-house action standards and control programs give us an edge only if we know what actions to take. Preserved products should show fast-acting bactericidal activity (e.g., 7 log reduction in 4 days). In the absence of such rapid kills, data could be generated showing that the package protects the product adequately during use. These requirements increase our chances for self-sterilization of product units during consumer use.

We should anticipate preservative failure, and identify at least two suitable preservative systems for each new product early in the development process. We can protect cosmetics from contamination during manufacturing by strict enforcement of sanitation. We should rigidly enforce cleanliness of process water, other raw materials, manufacturing processes, and personal hygiene. Our jobs will be increasingly more difficult in the future. Today's bacteria have declared war. We must continue developing weapons and honing our fighting skills to win that war in future battles.

REFERENCES

1. *Standard Methods for the Examination of Dairy Products,* 16th ed., American Public Health Association, Washington, D.C., 1993.
2. *Standard Methods for the Examination of Water and Waste Water,* American Public Health Association, Washington, D.C., 1992.
3. *Compendium of Methods for the Microbiological Examination of Foods,* 3rd ed., American Public Health Association, Washington, D.C., 1992.
4. *U.S. Pharmacopeia XXIII,* U.S. Pharmacopeial Convention, Rockville, MD, 1995.
5. *FDA Bacteriological Analytical Manual,* 7th ed., Association of Official Analytical Chemists, Washington, D.C., 1992.
6. *CTFA Microbiology Technical Guidelines,* Cosmetics Toiletries and Fragrance Association, Washington, D.C., 1993.

7. Farrington, J. K., Martz, E. L., Wells, S. J., Ennis, C. C., Holder, J., Levchuk, J. W., Avis, K. E., Hoffman, P. S., Hitchins, A. D., and Madden, J. M., Ability of laboratory methods to predict in-use efficacy of antimicrobial preservaticves in an experimental cosmetic, *Appl. Environ. Microbiol.*, 60, 4553, 1994.
8. Brannan, D. K., Dille, J. C., and Kaufman, D. J., Correlation of *in vitro* challenge testing with consumer-use testing for cosmetic products, *Appl. Environ. Microbiol.*, 53, 1827, 1987.
9. Lindstrom, S. M., Consumer use testing: Assurance of microbiological product safety, *Cosmet. Toilet.*, 101, 71, 1986.

2 BIOLOGY OF MICROBES

Daniel K. Brannan

CONTENTS

0-8493-3713-5/97/$0.00+$.50

GENERAL BIOLOGICAL CONCEPTS

The Philosophy of Biology

The purpose of this chapter is to present to the nonscientist the critical information needed to understand microbiology such that it does not seem to be such a black art. It is also provided to document what has gone on for so many years: anecdotal stories of how to prevent contamination and how to preserve cosmetics. Anecdotal stories are little more than factoids of questionable provenance which masquerade as science; perhaps this chapter and indeed this entire book can turn such anecdotal stories into data-based facts.

Perhaps the one critical attitude that escapes most microbiologists today is a feel for the organism. Often a microbiologist learns so many technical details that he or she does not have time to simply become familiar with a microbe by growing it and caring for it. Microbiology students should, at some point in their career, have experience in being a curator for a culture collection in order to learn "how to think like a bug." These experiences develop an empathy for the microbes that often allows one to intuitively reach decisions that are based on thousands of data points which are processed subconsciously due to a gestalt not developed in those who have never acquired such understanding. It is this empathy that separates the real microbiologist from just a technician. It starts with a belief that a common bond exists between humans and even the most bizarre of the microbes—a bond that allows us to appreciate the organism simply because it represents life itself, and a bond that makes us realize that we all have the same ultimate origin as a result of the same creative wisdom.

A real biologist is interested in studying the organism simply because it allows him or her to gain an understanding of the creature rather than its potential to be converted into money or prestige or the next cure for some disease to benefit human-kind. A real biologist studies life simply because he or she wants to understand the incredible design behind the adaptability that has enabled all of nature to develop into the various species we see today occupying every conceivable niche available. And, in the process, the real biologist is struck with an incredible awe of life itself. One individual who was very much filled with such awe of the creation around him was Darwin.[1] In his own words: "To my mind it accords better with what we know of the laws impressed on matter by the Creator, that the production and extinction of the past and present inhabitants of the world should have been due to secondary causes, like those determining the birth and death of the individual.... There is grandeur in this view of life, with its several powers having been originally breathed into a few forms or into one; and that, whilst this planet has gone cycling on according to the fixed law of gravity, from so simple a beginning endless forms most beautiful and most wonderful have been, and are being, evolved." Thus, the real biologist will experience an almost reverent awe when he or she studies "the laws impressed on matter by the Creator" in producing the "grandeur in this view of life" such that life is produced "from so simple a beginning [into] endless forms most beautiful and most wonderful." For in studying such laws that govern nature, one may just begin to glimpse the mind of who Darwin referred to as the Creator.

The cosmetic microbiologist must add to this appreciation of life and feel for the organism the details of technical training. He or she must balance a variety of factors to provide for safe, unspoiled quality products. In addition to knowing sanitation and preservatives, he or she must understand microbial physiology, pathogenic microbiology, and microbial ecology. In addition to microbiology, he or she must understand organic and physical chemistry, toxicology, engineering, manufacturing and processing, and regulatory/environmental law. The cosmetic microbiologist must use all this education and knowledge within the context of the business needs of the company and be able to balance risk/benefit to the consumer using the product. The microbiologist that best fits this description is the one who has a good solid liberal arts undergraduate education followed up with post graduate training in microbiology. Finally and most importantly, this person must have the highest of ethical standards. They must consider themselves as part of the cadre of health-care providers in the world dedicated to serving humankind via the mission of providing them with microbially safe and efficacious products.

Evolution and Adaptation

Dobzhansky once stated that nothing in biology makes sense outside the light of evolution. Perhaps this was slightly an overstatement unless one defines evolution as simply change, variation, and adaptation of organisms to various environments. This adaptation is easily seen in man's artificial selection of domesticated animals and can even be seen in nature's natural selection during one's lifetime by looking at the beak of the finch as the Grants have shown. Natural selection is observable, testable, and, when models are based on it, predictable. This is what makes evolutionary theory so rich: it transforms biology from a descriptive science to a predictive, explanatory science.

Certainly, as Stephen Gould points out, there have been many abuses of this explanatory ability in the many "just-so" Panglossian-like explanations for why nature is just the way it is. And certainly there are still many gaps to fill and more data gathered and better explanations offered before macroevolution (or the general theory of organic evolution) can advance beyond more than just a nonsupernatural metaphysics based on materialistic naturalism combined with randomness and blind chance. Even Darwin saw this and decided to reject a completely naturalistic explanation for origins. However, there have been just as many other instances where great insight and explanations have been provided to illustrate how a species adapts (microevolution) to changing environments.

Any microbiologist is well aware of the power of this latter type of biological evolution where we are constantly needing to develop new biocides and antibiotics at a frenetic pace just to keep one step ahead of microbial evolution. Anyone who believes in fixity of species in the microbial world needs only to work in an environment where the microbes adapt so rapidly to the next new biocide or antibiotic. They change so rapidly that even a minor type of punctuated equilibrium has been shown by Richard Lenski's lab and can be observed in *Escherichia coli* in a matter of years based on cell size increases.

One of the key areas to understand is the way that microorganisms evolve in order to become tolerant to biocides. A incredibly naive idea is that the tolerance mechanisms against biocides are similar to those mechanisms found in antibiotic resistance. Antibiotic resistance develops as a result of molecular changes at a single specific site of attack by the antibiotic (e.g., the P site of 50s ribosomes). Biocides attack at multiple sites (e.g., chlorine oxidizes all proteins, carbohydrates, and fats in a cell). Therefore, minor mutational events or acquisition of some plasmid containing a resistance gene that codes for some specific r-factor is an inappropriate model to describe biocide tolerance.

The development of tolerance for many biocides is more likely a result of phenotypic changes where the population shifts to higher production of capsule. This approach is a true Darwinian mechanism where the biocide forces a population bottleneck allowing only a few hardy cells to survive that pass on the genes for that survival to their offspring (e.g., genes that express more frequently for the production of capsule). This is analogous to the hairy-elephant-population-gets-hairier story during an advancing ice age. Or, the cells may begin a primitive form of communication as in quorum sensing to stimulate physical community developments within biofilms where certain organisms exist as protector guilds for other organisms.

Certainly, specific biochemical mechanisms (e.g., formaldehyde dehydrogenase) of resistance to biocides exist in the case of formaldehyde tolerance where it is degraded. However, tolerance to biocides that act at multiple sites must not be assumed for all microorganisms; in many cases, the tolerance claimed may be a result of physical factors such as capsule formation or existing within a clump of organisms or within a biofilm.

Regardless of the adapting mechanisms extant in a microbial population, the key idea to get across to nonbiologists is that nature is never stable. The microorganisms are changing and adapting constantly. They are not just little sacs of enzymes suspended in a liquid as the reductionist-minded analytical chemist thinks; they are not just another ingredient to be gotten rid of as the pragmatic process engineer thinks. They are clever, adapting little sparks of life containing the wisdom of the ages within them which allows them to survive no matter what we throw at them. The new biocides we develop today will not work tomorrow. The microbiologist will always have work to do to keep one step ahead of the ever-adapting, ever-changing, ever-evolving microbes.

THE BACTERIA

The bacteria play a major role in the world's ecosystems. They are so ubiquitous and important that they are considered by some to be the lifeblood of the planet. According to the Gaia hypothesis, they are involved in biogeochemical cycling to maintain Earth's homeostasis. They occupy every conceivable niche and some that we may not have even dreamed possible. Consequently, they are found throughout cosmetic and drug manufacturing plants. A single pin-head may contain well over 1 billion bacterial cells. A single handful of soil represents an entire universe of bacterial possibilities, all capable of adapting to even the harshest environments. Well-preserved products and scrupulously clean manufacturing environments help prevent these organisms from establishing niches within a consumer product. But their genetic adaptability and remarkable evolutionary capability presents a moving target that is difficult to control without constant surveillance. This chapter presents an introduction into microbiology in order to enable even the nonmicrobiologist to understand at least the basics of this very complex and ever changing world of microorganisms.

Physiology and Biochemistry of Microorganisms

Size, Shape, and Organization

Prokaryotic cells such as bacteria are without internal membrane-bound organelles. They are much simpler than eukaryotic cells such as the cells of our body that do contain internal membrane-bound organelles. Since bacteria are so simple, our first inclination to is believe that they should be simple in shape. In fact, most are either cylindrical (rod-shaped) or spherical (cocci). However, a variety of other shapes as well can occur: spirals (spirilla and spirochetes), curves (vibrios), even squares and triangles. The shape of the cell is determined by its cell wall. Bacteria are also smaller than most eukaryotes. Some bacteria are as small 100 nm (*Mycoplasma* spp.); some are as large as 60 × 800 μm (*Epulopisicum fishelsoni*). The large size of *E. fishelsoni* contradicts the small size assumption by overcoming the hypothetical diffusion limits through a plasma membrane that is highly convoluted.

Bacterial Organelles

A variety of structures exist within prokaryotes. Figure 2-1 and Table 2-1 describe a typical, generalized bacterial cell. Not all bacteria are identical or contain all structures. In fact, a major difference exists between Gram-negative and Gram-positive cells, particularly in their cell walls.

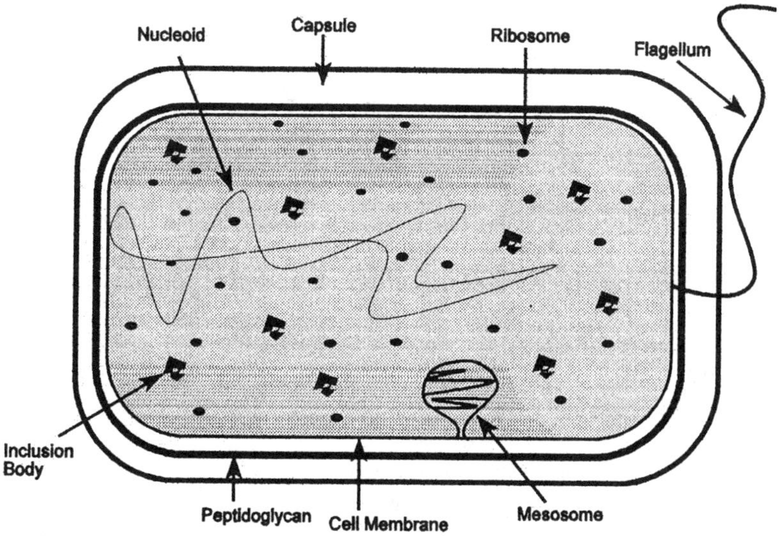

Figure 2-1 Typical bacterial cell. Shown is a Gram-positive cell.

TABLE 2-1

FUNCTIONS OF VARIOUS PROKARYOTIC CELL STRUCTURES

Structure	Function
Capsules/Slime Layers	Resistance to phagocytosis, adherence to surfaces
Fimbriae/Pili	Attachment to surfaces, bacterial conjugation
Flagella	Movement
Cell Wall	Provides shape and protection from lysis in dilute solutions
Plasma Membrane	Selectively permeable barrier and boundary of cell. Location of respiration and photosynthesis; receptors for chemotaxis
Periplasmic Space	Site of hydrolytic enzymes and proteins for nutrient uptake/processing
Ribosomes	Protein synthesis
Inclusion Bodies	Storage
Gas Vacuoles	Buoyancy for floating (some bacteria)
Nucleoid	Site of genetic material (e.g., DNA)
Endospores	Survival under harsh environments (some bacteria).

In general, a cell wall and plasma (or cytoplasmic) membrane surrounds prokaryotic cells. Some prokaryotes lack the cell wall but all have a cytoplasmic membrane. The membrane can invaginate to form mesosomes and other internal membranous structures within the cytoplasm. In most prokaryotic cells, however, the cytoplasm is very simple and without any membrane-bound internal organelles. This is in contrast to what is found in eukaryotic cells. The genetic material (e.g., DNA) is in a region called the nucleoid. This region is not surrounded by a nuclear membrane in a nucleus like it is in eukaryote. Ribosomes and inclusion bodies are found throughout the cytoplasm. One organelle that extends outside the cell wall is the flagellum. Around the cell wall is a capsule and pili or fimbriae. These structures help the bacterium stick to surfaces and form biofilms. They also allow the bacteria to stick to each other and clump especially when exposed to biocides or other adverse conditions.

Cell Wall and Cytoplasmic Membrane

Nearly all prokaryotes have cell walls. The only exceptions are the mycoplasmas and a few archaeobacteria. The cell wall gives the bacterium its strength and shape. Without the wall, the bacterium would lyse due to the effect of osmosis. Most bacteria exist in a dilute external environment. In fact, the environment inside the cell is more concentrated in solutes (dissolved materials) than the outside. Therefore, water flows into the cell continuously. Without the rigid cell wall, the bacterial cell would lyse. A whimsical analogy is water balloons. If the balloon is placed into a rigid box before one fills it with water, it cannot be filled so full that it pops. Outside the box, it will pop when filled too full.

Between the cell wall and cytoplasmic membrane is the periplasmic space. This is actually a gel-like area. It is filled with hydrolytic enzymes and binding proteins that digest nutrients and transport them into the cell.

Next comes the cytoplasmic membrane. Simplistically, the novice would claim that this structure is not all that different from any other cell membrane—eukaryotic or prokaryotic. After all, it has proteins and lipids. It surrounds the cytoplasm. It is the main point of contact between the cell and its environment. It is selectively permeable to allow only certain molecules to get into the cell. It transports those large molecules that cannot diffuse. It carries on metabolic functions such as respiration, photosynthesis, and some biosyntheses. It is composed of lipids that arrange into a bilayer (because of a lipid molecule's amphipathic nature—where one end is hydrophilic and the other is hydrophobic). However, the few differences that do exist between eukaryotic and prokaryotic membranes are significant. The bacterial membranes do not have cholesterol in them like our cells do. The proteins are different. So even though membranes have a common basic design, they differ widely in their various structural and functional capacities.

The model of a membrane that is widely accepted by most scientists is called the fluid mosaic model. One can think of this model as a layer of oil (lipid membrane) on top of an ocean (the cytoplasm) into which are interspersed ships of protein. Except in our model, there is dilute water on top of the oil layer (the external environment of the cell) as well. Like ships on the ocean, these proteins move laterally around the surface. The protein "ships" can be either peripheral (loosely connected to the membrane and dissolve in water) or integral (hard to extract from the membrane and insoluble in water).

Peptidoglycan. The cytoplasmic membrane of any cell is delicate and prone to rupture unless the cell exists in an isotonic state. This is why physiological saline is used, not distilled water, to replace body fluids so as not to rupture red blood cells. But bacteria rarely have the luxury of floating in an isotonic environment unless they are living inside our bloodstream. Most of the time, they find themselves in a hypotonic environment where water is pouring into the cell. Peptidoglycan makes up the structure of the cell wall to give rigidity and strength to it to combat the hypotonic environments.

Peptidoglycan (a.k.a. murein) is a net-like structure that makes up the cell wall. Peptidoglycan is a polymer made of two amino sugars (*N*-acetylglucosamine and *N*-acetylmuramic acid) and several amino acids (some unique only to prokaryotes and not even found in most proteins). Figure 2-2 shows the structure of peptidoglycan.

There are considerable differences between Gram-negative and Gram-positive cell walls. One of the major differences is the peptidoglycan. In Gram-positive cells, there is a pentaglycine bridge between the D-alanine and L-lysine of the tetrapeptide that comes off the *N*-acetylmuramic acid. In Gram-negative cells, there is a direct link between alanine and lysine.

The definition of Gram-positive and Gram-negative cell walls relates to two concepts: the color of the cell after the Gram stain and the physiological structure of the wall itself. When Christian Gram developed the gram stain, he developed a staining technique that allowed some bacteria to stain pink and others purple. At the time, he did not know why this occurred but, as a result of this differential staining, microbiologists began calling purple stained bacteria "positive" and pink stained bacteria "negative." Later on, they found out that the purple staining Gram-positive bacteria usually had thicker peptidoglycan layers and the pink staining Gram-negative

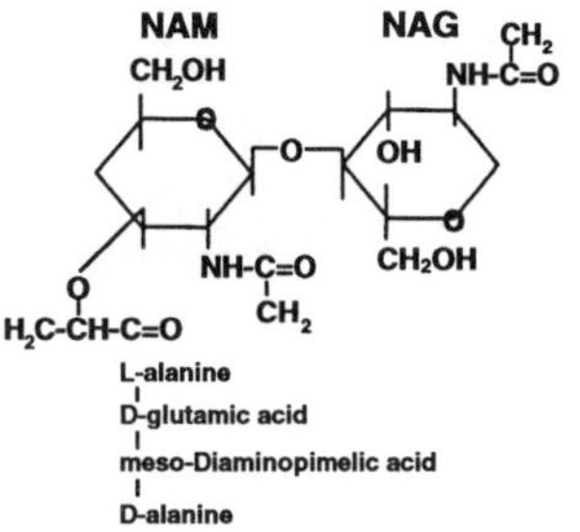

Figure 2-2 Peptidoglycan subunit where NAG is N-acetylglucosamine and NAM is N-acetyl muramic acid. Attached to the NAM is a tetrapeptide. In Gram negative peptidoglycan, the crosslink between two NAG-NAM strands is direct between d-alanine and DAP. In Gram positives, the crosslink is indirect via a penta-glycine bridge.

bacteria had thinner peptidoglycan layers. We will explore additional differences between Gram-postive and Gram-negative bacteria next.

Gram-positive Walls. Gram-positive bacteria have a very thick peptidoglycan layer. The peptidoglycan has the pentaglycine bridge between the tetrapeptide coming off the *N*-acetylmuramic acid (NAM). The NAM is polymerized to *N*-acetylglucosamine (NAG). Gram-positive cell walls also have teichoic acids in them (Figure 2-3). These are ribitol and glycerol phosphate polymers. Coming off the ribitol and glycerol may be amino acids or sugars. The teichoic acids attach to the peptidoglycan layer and extend to the outside of the cell where they give the cell a negative charge. They can even extend all the way down into the cell membrane and attach to lipids (lipoteichoic acids). They are like tie rods that hold the peptidoglycan to the membrane, but their true function has not been entirely clarified yet. They are not in Gram-negative cell walls.

Gram-negative Walls. Gram-negative bacteria have a thin peptidoglycan where the tetrapeptide coming off the NAM is directly linked to the one coming off an NAM on an adjacent strand of NAG-NAM polymer. The complexity of Gram-negative cell walls is astounding. There is an outer membrane on the outside of the thin peptidoglycan layer. Linking this outer membrane to the peptidoglycan is a lipoprotein called Braun's lipoprotein. It is covalently linked to the peptidoglycan with its hydrophobic end stuck in the lipids of the underlying surface of the outer membrane (Figure 2-4).

The outer membrane of Gram-negative bacteria is made of a lipid bilayer like most classical membranes. But its uniqueness comes from the lipopolysaccharides (LPS) that extend from the outer layer of the outer membrane into the environment. The lipopolysaccharides are very large molecules made of lipid and carbohydrate. A diagram of a LPS is shown in Figure 2-5. It is the LPS that elicits the most antibody response

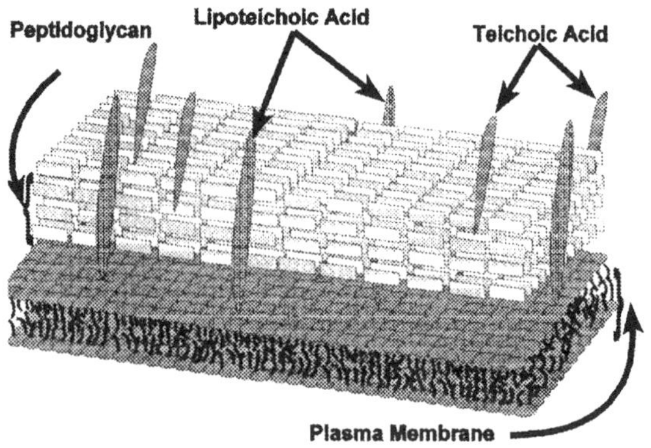

Figure 2-3 Gram-positive cell wall.

(to the O-region of the LPS) during an infection. It is also this region that rapidly changes to avoid antibody attack. LPS also confers a negative charge to the bacterial surface and it helps stabilize the outer membrane. This is also the component that acts as the endotoxin.

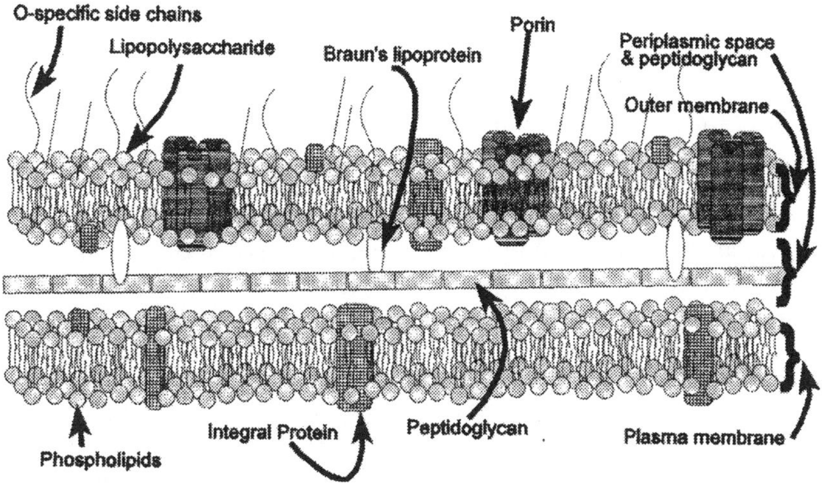

Figure 2-4 Gram-negative cell wall.

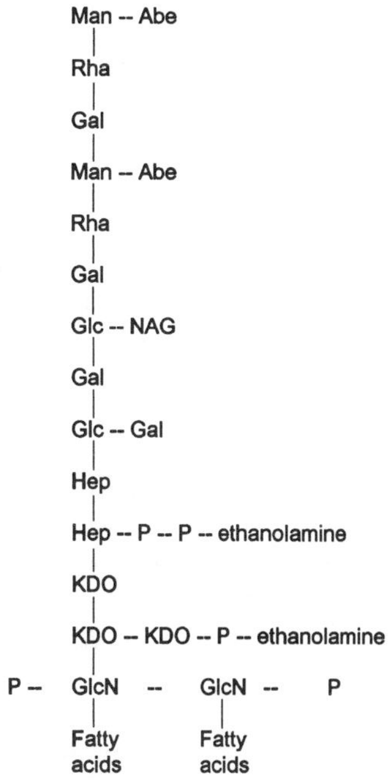

Figure 2-5 Lipopolysaccharide from *Salmonella.* Abbreviations: *Abe*, abequose; *Gal*, galactose; *Glc*, glucose; *GlcN*, glucosamine; *Hep*, heptulose; *KDO*, 2-keto-3-deoxy octonate; *Man*, mannose; *NAG*, N-acetylglucosamine; *P*, phosphate; *Rha*, L-rhamnose.

The outer membrane really does not act as a selectively permeable membrane in the same way that the cytoplasmic membrane does. Instead, it offers protection and slows entry of toxic substances into the cell. It does, however, allow small molecules like monosaccharides to pass to the cell membrane via porin proteins. These form a channel through which these smaller molecules (<700 daltons) pass. Large molecules are transported by specific carrier proteins. It has some selective permeability about it, but should never be thought of as another cytoplasmic membrane.

Gram Stain. When one does the Gram stain, he or she spreads a thin suspension of bacteria on a glass slide, lets it dry on the slide, and heats it gently to fix the bacteria onto the slide. The first step is to put a crystal violet solution on the slide for about a minute and rinse with water. The bacteria absorb the dye and turn purple. The next step is to put Gram's iodine on the slide. This is a mordant that "sets" the purple dye. Next, one rinses with an acetone-alcohol mixture to try to remove the set dye. Finally, one counterstains with saffranin, a pink dye. If the crystal violet dye cannot be removed with the acetone-alcohol, then the saffranin counterstain is not even seen; the bacterium is so dark purple that the pink dye does not contribute to the color. If the crystal violet does get removed, then the bacteria are colorless until the pink saffranin counterstain is added. Bacteria that stain purple are called Gram-positive (they were positive for keeping the first dye). The ones that stained pink are called Gram-negative.

The key to understanding how the Gram stain works is in the cell wall. One hypothesis is that the alcohol shrinks the molecular pores of the thick peptidoglycan of Gram-positive cells where the crystal violet is trapped. In the Gram-negative cells, the thin peptidoglycan is not highly cross-linked. So the pores are bigger and more permeable, allowing the crystal violet to leak out quicker when one decolorizes. The alcohol also dissolves the lipid of the outer membrane to allow the stain to escape.

Cytoplasm, Mesosomes, Ribosomes, and other Inclusions

A bacterial cell minus its wall is a protoplast. A protoplast includes the plasma membrane, the cytoplasm, and everything within it. The prokaryotic cytoplasm, however, does not have typical unit membrane-bound internal organelles. Within the cytoplasm is the nucleoid where the genetic material, DNA, is localized. Also, within the cytoplasm are the enzymes needed for growth and metabolism, the machinery for manufacturing those enzymes (ribosomes), and some internal membrane structures called mesosomes. Mesosomes are actually invaginations of the plasma membrane. Finally, some bacteria also contain inclusion bodies made up of polyphosphate, cyanophycin, and glycogen. These inclusions are not usually membrane-bound. Other bacteria have inclusions bound by a single-layered nonunit membrane. These are made up of poly-β-hydroxybutyrate, sulfur, carboxysomes and gas vacuoles.

Mesosomes are rather enigmatic. Their exact function remains uncertain despite finding them in Gram-positive and Gram-negative bacteria. They are often found close to the septa of dividing cells; sometimes they even appear to be attached to the chromosome. Perhaps they are involved in cell wall formation. Perhaps they play a role in chromosome separation during cell division being somewhat analogous to spindle fibers in eukaryotic cell mitosis. Maybe they are involved in secretory processes. Some microbiologists dismiss them entirely, claiming they are artifacts of the fixation process used in preparing samples for electron microscopy. Since we also see them in freeze etches where little or no artifacts develop from chemical fixation, it is doubtful that this skepticism will hold. Other, more complex, membrane systems of invaginated plasma membrane are seen in the complex and extensive foldings within the cyanobacteria, nitrifying bacteria, and purple bacteria. The current thinking is that these organisms require the larger surface area that these infoldings provide in order to carry on greater metabolic activity.

Ribosomes are packed into the cytoplasm; some are loosely attached to the plasma membrane. They are made of both ribonucleic acid and protein. They are where RNA is translated into protein after the RNA is transcribed from DNA. In bacterial systems (and in mitochondria) they are composed of two components: a 50S and a 30S subunit (the S standing for Svedberg unit, a measure of sedimentation based on the particle's volume, shape, and weight). Together, these two subunits make up a single ribosome and now weigh 70S (it is supposed to not add up since the Svedberg unit is based on more than just weight alone).

The Nucleoid

This structure is not surrounded by a membrane either. Eukaryotes always have their genetic material within a nucleus surrounded by a membrane; they have two or more linear chromosomes. Prokaryotes always have only one circular chromosome located within a region called a nucleoid. The nucleoid is apparently attached to the cell membrane which may be involved in cell division via the aid of mesosomes.

Bacteria also contain plasmids. These are extrachromosomal pieces of DNA (also circular) that replicate independently of the chromosome. They can also be incorporated into the chromosome. They rarely have genes that are absolutely required by the organism for growth and metabolism. But they often carry genes that are very useful for survival: resistance genes that make them able to withstand antibiotics, genes that allow the organism to produce a toxin, or genes that give the bacterium some other selective advantage.

Organelles Outside the Wall

Flagella, Fimbriae, and Pili. Fimbriae (or pili) are thin protein hairs on the outer surface of a Gram-negative bacterium that cause the bacteria to stick on surfaces. If it were not for these pili, bacteria would not be as able to cause infection by attaching to host tissue. They would not be as able to form biofilms on pipes of water systems to cause a reservoir of contamination or endotoxin. They would not be able to attach to ship hulls leading to further corrosion. Some of the pili are even involved in providing a passage for DNA from one bacterium to another during conjugation.

Flagella allow bacteria to move. Unlike eukaryotic flagella, the prokaryotic flagellum rotates rather than moving from side to side. Also, it is a relatively simple thread-like appendage of protein extending from the plasma membrane and cell wall (about 20 μm long and 20 nm thick). This portion is the filament. Compared to the 9 + 2 filament arrangement of eukaryotic flagella, this structure is simple. A little more complexity occurs when we look at the bacterial flagellum's basal body. It is embedded within the bacterial cell wall. It consists of a number of rings which vary depending on whether the bacterium is Gram-negative or Gram-positive. A hook links the filament and the basal body. The basic difference in Gram-negative and Gram-positive flagella is in the number of basal body rings. Gram-negative bacteria usually have four basal body rings. The first two rings are attached to the outer membrane (L-ring) and peptidoglycan layer (P-ring). The inner two rings contact the periplasmic space (S-ring) and the plasma membrane (M-ring). In the Gram-positives, only two rings exist with one attached to the plasma membrane and the other attached to the peptidoglycan.

Capsules and Slime Layers. Outside the cell wall, a bacterium may have a layer of material that can be fairly well organized and not easily washed away. This is the capsule. If the material is easily washed off and not really organized, it is a slime layer. Many scientists prefer to not worry about these distinctions and so refer to both as the glycocalyx. In either case, they are made of polysaccharides or poly-amino acid (in some bacteria) surrounding the outer cell wall of bacteria. The capsule helps the bacterium resist phagocytosis when it infects a host. Encapsulated bacterial pathogens are usually much more virulent than those varieties without a capsule. The capsule also helps the bacterium avoid drying, aids it in attachment to surfaces as a biofilm, helps it avoid predation from zooplankton, and protects it from detergent and biocide action. From a practical standpoint, encapsulated bacteria are more likely to develop resistance to preservatives and biocides in manufacturing conditions.

Metabolism

In order to understand metabolism fully, the student needs to understand at least the basics of thermodynamics, chemical reactions coming to equilibrium and oxidation-reduction reactions. In addition, one should understand the role of enzymes in biochemical reactions, and the central role of ATP in mediating the flow of energy from one process to the next. This section will provide only the rudimentary information to accomplish an understanding of these points.

All organisms require energy to do work. This work comes in three forms: (1) chemical work where complex biological molecules are synthesized from simpler molecules; (2) transport work where molecules are brought into or taken out of the cell; and (3) mechanical work where energy is used in movement by the organism or where the structures of the organism change position. Most ecosystems receive their energy to do work from sunlight and through photosynthesis. A few ecosystems receive their energy from geothermal energy and through chemosynthesis using inorganic energy sources such as sulfide (e.g., oceanic thermal vents). The light energy from the Sun (e.g., photosynthesis) and the geothermal energy (sulfide) from Earth (e.g., chemosynthesis) is trapped in the form of chemical energy stored in complex compounds. The organisms involved in this primary production of chemical energy from light or heat/inorganic compounds (producers) serve as energy for chemoheterotrophs (consumers) which use complex organic molecules as a source of material and energy for building their own cellular structures.

In transforming light or chemicals into usable energy for cellular processes, cells use adenosine triphosphate (ATP) as an energy exchange molecule to make free energy available to do work. ATP is a high-energy molecule that is energetic due to the repulsive forces that are overcome by covalent bonds between the highly charged triphosphates. At pH 7.0, the linear triphosphates are completely ionized giving the ATP molecule four negative charges. These negative charges are very close to each other and repel each other. If the covalent bond that overcomes these charges is hydrolyzed, this electrostatic stress between the phosphates is relieved and the difference in the energy of the products (ATP and water) and the reactants (ADP and inorganic phosphate) releases sufficient energy for use in coupling to another reaction. When ATP is hydrolyzed, the reaction releases a $\Delta G^{O\prime}$ of –7.3 kcal/mole (free energy available for work).

Energy flow may be described in terms of chemical reactions that proceed to equilibrium such as we described for ATP hydrolysis. However, free energy changes are also involved in oxidation-reduction reactions. These reactions occur when electrons move from a donor to an acceptor. Each process is described using two slightly different thermodynamic equations:

$$\text{Chemical Energy: } \Delta G^{O\prime} = -RT\,(\ln K_{eq}), \quad (1)$$

where R is the gas constant (1.9872 cal/mole-degree), T is the absolute temperature of the reaction, and K_{eq} is the equilibrium constant for the particular reaction, and

$$\text{Redox Energy: } \Delta G^{O\prime} = -nF\,(\Delta E_O'), \quad (2)$$

where n is the number of electrons transferred, F is the Faraday constant (23,062 cal/mole-volt), and $\Delta E'_0$ is the difference between the reduction potentials of the coupled redox reaction (the chemical that is oxidized to give electrons to the chemical that accepts them getting reduced is the "coupled redox reaction"). Chemical energy is what is involved during substrate-level phosphorylations such as the step from 1,3 bisphosphoglycerate to 3-phosphoglycerate during glycolysis. Redox energy is what is involved during the production of reduced NAD produced in the Krebs cycle. Subsequently, these electron carriers donate their electrons to a cytochrome chain where energy is produced by creating a hydrogen ion gradient across the cell membrane that permits the production of ATP via ATP synthase as hydrogen ions flow back into the cell.

None of these processes would work effectively without the presence of enzymes that catalyze the reactions. Enzymes are simply protein catalysts that are specific for making a chemical reaction go from the reactants to the products. Enzymes do not change the equilibrium of the reaction nor do they cause more products to form. As catalysts, the enzymes simply speed up the reaction to its final equilibrium. They do this by lowering the activation energy required to bring the reacting molecules together in the correct way so they can reach what is called a transition-state complex which then can decompose into the products, assuming the energy state of the products is lower than that of the reactants. How enzymes bring the reactants together to form the transition-state is an area of considerable research. Enzymes bring substrates together at an active site where the activation is lowered because far more molecules of the reactants are concentrated in the area for the reaction to take place. However, this concentration alone does not explain the whole process. In addition, the molecules are oriented into the correct position for the molecules to react to form the transition-state complex.

Carbohydrate Metabolism

Metabolism refers to all the chemical reactions going on in a cell to allow the production of energy (catabolism) and then use that energy to allow for synthesis of complex molecules to create the ordered cell or organism (anabolism). Catabolism involves breakdown (oxidation) of organic compounds to provide energy. Catabolism can also involve the oxidation of inorganic compounds (chemolithotrophy) or the use of light (photosynthesis) to provide energy.

For heterotrophs, catabolism can easily be conceptualized as three interrelated stages. The first stage breaks down large molecules and polymers such as carbohydrates, proteins, and lipids. The chemical process is usually via hydrolysis. Carbohydrates are broken down into monosaccharides, proteins into amino acids, and lipids into glycerol and fatty acids. During the second stage, these are further broken down into even simpler molecules such as acetyl CoA and pyruvate (during monosaccharide, glycerol, and fatty acid breakdown), and into tricarboxylic acid cycle intermediates (during amino acid breakdown). During this second stage, the production of ATP and electron carriers such as NADH and $FADH_2$ occurs. Finally, these simple molecules are completely oxidized into CO_2 and more ATP, NADH, and $FADH_2$ are produced. The electron carriers are further oxidized via the electron transport chain to yield large amounts of ATP, producing water and carbon dioxide in the process.

What is fascinating about these reactions is that they all converge into three similar and common catabolic pathways: glycolysis, the tricarboxylic acid cycle, and the electron transport chain. What is even more fascinating is that these pathways exist as amphibolic pathways. This means that they can function both ways: anabolically as well as catabolically. Most of the reactions in glycolysis and the tricarboxylic acid cycle are reversible. Thus the molecules produced along the pathways can also be used as precursors for synthesizing macromolecules needed by the cell for growth and repair. There are a few irreversible catabolic steps that use special enzymes to catalyze the reverse reaction (i.e., phosphofructokinase) in order to permit regulation of whether the amphibolic pathway will function as a catalytic or an anabolic pathway.

Glycolysis. Glycolysis or the Embden-Meyerhof pathway is one of three major pathways to break down sugar to pyruvate. The other major pathways include the Entner-Doudoroff and the pentose phosphate pathway. Glycolysis is considered the more common of the pathways since it is found in all major groups of microorganisms and functions regardless of the presence of oxygen. It is usually divided into two main parts to better conceptualize how it works. The first part is referred to as the 6-carbon sugar stage. This stage involves two phosphorylating steps to convert glucose to fructose-1,6-bisphosphate. Two molecules of ATP are used to prime this step. The second step is the 3-carbon sugar stage which cleaves the fructose-1,6 bisphosphate into two 3-carbon molecules. These are each converted to pyruvate forming a total of two pyruvate molecules, two NADH molecules (an electron carrier), and four ATP molecules. Since two ATP molecules are used to prime the reaction in step 1, we get a net of two ATP out of the process. The ATP is formed by the process known as substrate-level phosphorylation since ADP phosphorylation is coupled with the exergonic breakdown of a high-energy substrate molecule.

Pentose Phosphate Pathway. The pentose phosphate pathway (also known as the hexose monophosphate pathway) uses a different set of reactions to form 3-, 4-, 5-, 6-, and 7-carbon sugar phosphates. These are used to produce ATP and NADPH (used in biosynthesis). The pathway is primarily used to provide the carbon skeletons for the synthesis of amino acids, nucleic acids, and other macromolecules. If biosynthesis is not needed, the NADPH will be converted to NADH to feed the electron transport chain for the production of more ATP. A unique reaction in this pathway is the production of 3-ribulose-5-phosphate from 6-phosphogluconate.

Entner-Douderoff Pathway. This pathway also produces pyruvate like glycolysis, but it produces a lower yield of only one ATP, one NADPH, and one NADH. The key intermediate unique to this pathway is 2-keto-3-deoxy-6-phosphogluconate derived from 6-phosphogluconate.

Tricarboxylic Acid Cycle. The tricarboxylic acid (TCA) cycle starts after acetyl-CoA is produced from pyruvate. The steps involved in producing acetyl-CoA from pyruvate are a system of multienzymes known as the pyruvate dehydrogenase complex. In this process, just before the TCA cycle, pyruvate is coupled with coenzyme A after forming acetate from pyruvate by removing one of the carbon atoms as carbon dioxide. Thus, the acetyl-CoA formed is composed of coenzyme A bound to acetic acid via a high energy thiol ester bond.

The acetyl-CoA formed then reacts with the 4-carbon compound oxaloacetate to produce the 6-carbon molecule, citrate. In regenerating oxaloacetate from citrate, two carbon dioxide molecules are removed. The one ATP produced is from substrate phosphorylation when succinyl-CoA goes to succinate during part of the cycle. Also produced is an $FADH_2$, and 3 NADHs for each acetyl-CoA molecule that enters the cycle. These electron carriers can then enter the electron transport chain to produce more ATP as described further below (Electron Transport and Oxidation Phosphorylation).

Fermentations. In the absence of oxygen, the product of glycolysis, pyruvate, cannot be completely oxidized to carbon dioxide and water via the Krebs cycle to produce the copious quantities of NADH that get oxidized via the electron transport chain since no terminal electron acceptor (e.g., oxygen) is available. In fact, the NADH produced during glycolysis must be oxidized in order to have more NAD+ available for reuse during glycolysis. Basically, all fermentations are reactions that regenerate NAD+ from NADH when an electron acceptor for the electron transport chain is unavailable. The usual mechanism for regenerating NAD+ is to oxidize the NADH produced using pyruvate as an organic electron acceptor. Thus, instead of completely oxidizing pyruvate, it becomes reduced in order to make NAD+ available to keep glycolysis operating. The goal is to keep up a meager production of ATP when oxygen is absent (only two ATP per glucose molecule) and to keep up the production of intermediates for anabolism.

When pyruvate acts as the electron acceptor, it may experience several fates. It may be decarboxylated into acetaldehyde which then accepts NADH to form ethanol being catalyzed by alcohol dehydrogenase. Or, pyruvate may be directly reduced to lactic acid using lactate dehydrogenase to catalyze the oxidation of NADH. Since only lactic acid is formed, this is called homolactic fermentation. When several other products in addition to lactic acid are formed including ethanol and carbon dioxide, it is known as heterolactic fermentation. Lactic acid fermentors include various *Lactobacillus* spp. and some *Streptococcus* spp. Pyruvate may also be reduced to form formic acid, ethanol, and a variety of other acids (e.g., acetic, lactic, succinic). In this case, it is called a mixed acid fermentation. This is typical of many enteric bacteria such as *Escherichia, Salmonella,* and *Proteus.* If formate hydrogenlyase is present, the formate will be further converted to hydrogen and carbon dioxide. Alternatively, the pyruvate may be converted to acetoin which is then reduced to butanediol using NADH. Organisms that do this fermentation are *Enterobacter, Serratia,* and *Bacillus* spp. These processes serve to help identify the various organisms that perform them and can be used as diagnostic tools. Other fermentations may take place as well including the formation of propionic acid (*Propionibacterium*), isopropanol, butanol, and butyrate (*Clostridium*). These latter fermentations may be capitalized upon for formation of foods and fuel.

Protein Metabolism

Many of the spoilage organisms and pathogens are efficient protein degraders. They secrete protease exo-enzymes to hydrolyze proteins into amino acids which can be transported into the cell. Once the amino acids are brought into the cell, they can be deaminated to remove the amino group and produce tricarboxylic acid intermediates from which either intermediates for biosynthesis or energy may be obtained from the NADH produced which is coupled to electron transport.

Fat Metabolism

Lipids are common energy sources for microbes. These exist as triglycerides or phospholipids or other complex esters. They are hydrolyzed to simpler compounds such as glycerol and fatty acids by lipases. Many of the fatty acids are then catabolized by β-oxidation which produces acetyl Co-A which goes into the tricarboxylic acid cycle. Considerable energy and reducing power can be generated with each turn of the β-oxidation pathway.

Electron Transport and Oxidative Phosphorylation

The key to generating considerable energy is to use the reducing power of NADH and FADH generated from the metabolic pathways outlined above. This is accomplished through a series of redox reactions by passing the electrons from NADH and FADH via several electron carriers (e.g., cytochromes) to a final or terminal electron acceptor. This process occurs in the bacterial plasma membrane, whereas in eukaryotes it occurs in the inner mitochondrial membrane. During this process of transporting electrons along cytochromes, a considerable amount of energy is produced which is used to pump protons to the outside of the cell. This sets up a gradient with protons on one side of the membrane and electrons on the other. When the protons diffuse back into the cell to equalize the gradient, they flow back through an enzyme complex that includes ATP synthase which synthesizes ATP as a result of the energy from this proton-motive force. An alternative hypothesis has gained considerable force: the conformational change hypothesis. This idea is that the energy released during electron transport causes the ATP synthesizing enzyme to change conformation such that it can bind ADP and phosphate more efficiently and thus catalyze the reaction to ATP better. The whole process of coupling redox reactions to the formation of ATP by the mechanism of an electron transport chain is known as oxidative phosphorylation. This is contrasted with substrate-level phosphorylation which occurs in glycolysis.

Anaerobic Respiration

Many students of microbiology consider anaerobic processes to be synonymous with fermentation. In fact, many anaerobic processes still may have ATP formed as a result of oxidative phosphorylations if they only replace the terminal electron acceptor oxygen with an inorganic form of oxygen. For example, nitrate (NO_3^-), sulfate (SO_4^-), and carbonate (CO_3^-) all serve as an inorganic salt of oxygen in order for anaerobic respiration (the use of a cytochrome chain) to take place. The term respiration in microbiology does not mean that the microbe breathes air. Instead, it means that it has a cytochrome chain capable of oxidative phosphorylation. In the case of anaerobic respiration, we use inorganic salts of oxygen rather than oxygen itself as the terminal electron acceptor.

Oxidation of Inorganic Molecules

The organisms that gain energy using inorganic molecules such as sulfide, iron, ammonia, and hydrogen do not contaminate consumer products. These are known as chemolithotrophs and usually exist in extreme environments. Since they do not use organic molecules for energy, a carbon rich environment such as a consumer product is not an environment in which these organisms grow.

Biosynthesis

Our discussion of biosynthesis will be limited to heterotrophic organisms. We will not include photosynthesis and chemosynthesis (fixation of carbon dioxide), since these autotrophs would not be expected to contribute significantly to consumer product contamination.

Synthesis of Carbohydrates and Polysaccharides

The synthesis of glucose rather than its breakdown is done via the process of gluconeogenesis, which is essentially a reversal of glycolysis. There are only three of the ten steps that are different and require separate enzymes to catalyze the reverse reaction. Otherwise, the same seven steps that permitted glycolysis are simply reversed. In addition, fructose is made using the same pathway. These two sugars are the foundation for many other sugars and for phosphorylated derivatives of those sugars. Polysaccharides are then made from the phosphorylated derivatives of sugar nucleosides. These are then used to make cell walls of bacteria.

Synthesis of Purines, Pyrimidines, and Nucleic Acids

Since purines and pyrimidines make up the DNA and RNA of the cell, they are very critical for survival and reproduction. These are also involved in the formation of phosphorylated compounds used for energy like ATP.

Purines and Pyrimidines. Purine (adenosine and guanosine) biosynthesis involves eleven steps, seven other cofactors, and folic acid. The requirement for folic acid allows us to capitalize on sulfonamides as antibiotics since they block the bacteria being able to synthesize folic acid. We get all our folic acid from our food and can absorb it directly, whereas the bacteria must manufacture it for themselves. The first few steps involve production of inosinic acid being formed onto the ribose-5-phosphate sugar. Once this structure has been formed, the steps to make adenosine and guanosine monophosphate and then to phosphorylate them into triphosphates are fairly simple involving vitamin B_{12} as a cofactor and thioredoxin as a reducing agent.

Pyrimidine biosynthesis begins reacting aspartic acid (an amino acid) with carbamoyl phosphate (a high energy molecule formed from CO_2 and glutamine) to form the first pyrimidine product orotic acid. Then the ribose sugar is added. A simple decarboxylation of the orotidine monophosphate results in uridine which can be phosphorylated to UTP; a transamination of the UTP results in CTP. If the ribose of the UTP is decarboxylated and the uridine is methylated with the use of folic acid, we form deoxythymidine monophosphate.

DNA Replication. Now that we have the basic building blocks of RNA (UTP, CTP, ATP, GTP) and DNA (dTTP, dCTP, dATP, dGTP), we can begin DNA replication. RNA synthesis will be covered in the section below. DNA is a polymer of the four nucleotides: dTTP, dCTP, dATP, and dGTP (the "d" stands for deoxy, signifying that one less hydroxyl than is found in RNA is gone from the sugar portion of the nucleotide). As DNA is made, two complementary strands are made which coil around each other in a double helix form. The easiest way to visualize them is as a ladder with either side of the railings representing the two deoxyribose-phosphate strands and the steps as crosslinks of purines and pyrimidines between the two strands. The ladder, however, is twisted or coiled into a spiral. Bacteria carry enough information in their DNA to equal about 40 novels of 400 pages of words. A single human cell carries enough information in its DNA to equate to about 40,000 novels.

When DNA replication begins, the double strand comes apart. A useful metaphor to help envision this process is the unzipping of a zipper. Once the strands are "unzipped," free floating nucleotides present in the cell match with their complements on each side of the two strands. Thus, a new strand is formed along each of the open strands and a single DNA molecule becomes two.

To understand the process in detail requires far more effort. In reality, there are at least seven enzymes involved in carrying out the process just described in the paragraph above. We have an initiator protein, helicase, polymerases, repair nucleases, topoisomerase, single-strand DNA-binding proteins, and DNA ligase. The initiator protein first finds the right place to begin copying and guides the helicase to the correct position (an origin of replication site) on the nucleic acid. The helicase separates the DNA by breaking the weak bonds between the nucleotides to unwind the two strands of DNA. Then the polymerases arrive to join the free nucleotides to their matching complements on the old strands using the phosphate bond energy from the nucleotide to help form the new bond to the other nucleotides as they are added to the existing chain. These polymerases work along with primases which first synthesize a short (1 to 5 nucleotides long) RNA primer. This primer allows DNA polymerase to begin catalyzing the addition of nucleotides to a new strand complementary to the existing template upon which the new DNA synthesis is based. Since DNA replication must follow a specific direction and since the two strands of DNA are antiparallel, there are two polymerases: one is continuously adding new nucleotides to what is called the leading strand as the DNA separates due to the helicase. The other strand is called the lagging strand; it runs in the opposite direction of the leading strand. However, the polymerase attached to the lagging strand cannot advance in the opposite direction but must follow in the same direction as the polymerase attached to the leading strand. As a result, the solution to this apparent dilemma is to have the polymerase make a loop in the lagging strand and add nucleotides along the bottom half of it. As the lagging strand polymerase finishes a length, it drops the completed end and attaches to a new loop to continue linking nucleotides along a new stretch of DNA in a stop and start fashion. This leaves a series of completed double helix fragments of DNA called Okazaki fragments (named after the discover), separated by incomplete gaps along the DNA strand. These gaps are filled in by the enzyme DNA ligase. Meanwhile, the leading strand is made continuously. A variety of single-stranded DNA binding proteins keeps the single strands apart until the polymerase and ligase can complete the addition of complementary nucleotides. The topoisomerases relieve the tension on the double helix in advance of the gyrase. Repair nucleases recognize errors in replication and remove incorrect nucleotides to allow the polymerases and ligases to replace them with the correct nucleotide.

RNA Synthesis (Transcription). Information storage is useless unless some mechanism is available to act on the information. The cell has such a mechanism. This mechanism can be understood by using the manufacturing metaphor of the use of the Standard Operating Procedures manual. Imagine someone wanting to sanitize a large piece of equipment. His first step is to go to the SOP manual (i.e., DNA) to get the information on how to do the sanitization. He copies down the information and carries the copied information out to the factory floor rather than carry out the entire book or even the original page on sanitization. This is just the first step, of course. The information carried out to the factory floor must be acted upon. But this step will be discussed when we cover synthesis of amino acids and protein below. The process of copying information from DNA to RNA is known as transcription.

In transcription, many of the same mechanics of DNA replication exist. The DNA double helix is opened up and a new nucleotide chain is made, for example. But many significant differences exist. In transcription, only one of the DNA strands is used as a template; it is called the sense strand. Only a few genes are copied at a time, not the entire genome as in replication. The most significant difference, however, is that the nucleic acid made is not DNA but RNA (ribonucleic acid). This compound is basically the same as DNA but it has a different sugar, ribose, rather than deoxyribose, and it uses the nitrogenous base uracil rather than thymine. Thus, the process involves the enzyme RNA polymerase. This is a versatile enzyme that finds the starting point of the gene (the Pribnov box), opens the double helix up, copies it, and then closes it, releasing the RNA (called messenger RNA or mRNA) to be acted upon in the next step of protein synthesis or translation. It knows when to stop transcribing because there is also an ending point of the gene carrying a sequence of that code for the mRNA to form a hairpin loop that causes the polymerase to stop transcribing. Transcription also allows for the production of two other types of RNA polymers: transfer RNA (tRNA) and ribosomal RNA (rRNA). Transfer RNA carries amino acids during protein synthesis. Ribosomal RNA is the stuff of which ribosomes are made (along with protein); ribosomes are critical in synthesizing protein.

There are a few differences between how eukaryotes and prokaryotes transcribe DNA to RNA. For the most part, these are limited to the posttranscriptional modification. In prokaryotes, no additional modification of the message is needed (except that in some cyanobacteria and archeobacteria there is some posttranscriptional modification). But in eukaryotes, a significant amount of mRNA is removed and not expressed since these are intervening sequences (introns); the sequences that are expressed (exons) are spliced together to go out of the nucleus as a single mRNA molecule made up of several exons.

Synthesis of Amino Acids and Protein (translation)

The final act in expressing the genes of DNA is to translate the transcribed message where the cell makes sense (or translates) from the mRNA. This process involves ribosomes to translate the information in the mRNA into proteins (polymers of amino acids). These strings of amino acids twist and turn and fall into shape on their own just because of their sequence to form the enzymes and other proteinaceous structural components of the cell. The first question to address, though, is where do the amino acids themselves come from?

Amino Acid Synthesis. Simply put, the amino acids come from the digested polymers ingested by the cells. They can come from protein catabolism or from the intermediates of carbohydrate catabolism which are diverted from the catabolic process into the anabolic process of making amino acids. The making of amino acids requires energy in the form of ATP and so these synthetic processes only occur when there is plenty of ATP available from catabolism and the intermediates can be diverted away from the process of energy generation.

The first step in synthesizing amino acids is to assimilate nitrogen. One mechanism to do this is nitrogen fixation. This process occurs predominately in ecologically related bacteria such as *Rhizobium, Azotobacter,* and the cyanobacteria. It also occurs in some *Klebsiella* and *Clostridium.* These latter two are sometimes contaminants of personal care products, so we will touch briefly on nitrogen fixation. Realize, however, that nitrogen fixation is predominately a function of organisms that are not of concern to the cosmetic or drug microbiologist. Nitrogen fixation is the reduction of atmospheric gaseous nitrogen into ammonia by nitrogenase. It requires at least 6 electrons for reducing power and 12 ATP molecules.

Other more common mechanisms for assimilating nitrogen is by ammonia incorporation or assimilatory nitrate reduction. Ammonia is easily incorporated into amino acids by forming the amino acid alanine directly by amination of pyruvate using the enzyme alanine dehydrogenase. Alternatively, the cell can form glutamate (an amino acid) by aminating α-ketoglutarate (a TCA cycle intermediate) using the enzyme glutamate dehydrogenase. Once these two amino acids have been formed, the ammonia they carry (now called an α-amino group) can be transferred to other carbon skeletons of other catabolic intermediates by transaminations to form several other amino acids. Other mechanisms of assimilating nitrogen include use of two enzymes in tandem (glutamine synthetase and glutamate synthetase with the same net result of the formation of glutamate from α-ketoglutarate. Nitrate can also be assimilated by forming ammonia using the enzyme nitrate reductase and electrons from $NADPH+H^+$; the ammonia formed is then assimilated using the processes described above.

Most of the bacteria assimilate nitrogen by the processes described above. Once assimilated, it is then transformed into the 20 amino acids that make up the proteins life uses. Other microbes assimilate nitrogen already in the form of amino acids by the breakdown of proteins. The details of amino acid biosynthesis from basic carbon skeletons can be found in introductory biochemistry textbooks. We will limit the discussion here to cover the general concepts of amino acid biosynthesis. We have already highlighted how two amino acids (alanine and glutamate) are formed from ammination of pyruvate and α-ketoglutarate. Simply put, the rest of the amino acids are derived from similar processes but start with other intermediates of glycolysis and the TCA cycle. For example, aspartate is formed by transamination of oxaloacetate. Lysine, threonine, isoleucine, and methionine are then synthesized from modification of aspartate. Phenylalanine, tyrosine, and tryptophan are formed from intermediates of the pentose phosphate pathway. Glutamine, proline, and arginine are formed by modifying glutamate. Serine, glycine, and cysteine are formed from 3-phosphoglycerate. Valine and leucine are from pyruvate and acetyl-CoA.

Ribosome and tRNA Involvement. Now that we have amino acids made, we need to put them together into proteins. This step is the actual step of translation. It is carried out by ribosomes reading the mRNA (translating it) and involving tRNA (transfer RNA) that brings the amino acids to the ribosome (composed of ribosomal RNA and protein). Before going further, however, we need to understand the nature of the triplet code of nucleic acids that codes for each of the 20 amino acids required to make a protein.

The DNA molecule is composed of four nucleotides composed of a nitrogenous base and a sugar. The bases that make each of the four nucleotides different are guanine, cytosine, thymine, and adenine. When mRNA is made, the sequence of nucleotides is transcribed into the bases that pair with the DNA message: adenine pairs with thymine but in RNA uracil is used instead; thymine pairs with adenine; guanine pairs with cytosine, and vice versa. Each of the 20 amino acids are represented by a three letter code of nucleotides. For example, AUG means methionine. But the code is degenerate in the sense that in the third position, the nucleic acid can be several choices with the result the same. Thus, for example CCC, CCU, CCA, and CCG all code for proline. This sloppy base pairing is referred to as wobble and helps in decreasing the ill effects of DNA mutations. These same codes are also present in tRNA where specific tRNAs carry specific amino acids.

Protein synthesis is the final step in gene expression. In the following process, known as translation, the mRNA nucleotide sequence is translated into the amino acid sequence of a protein. The first stage is amino acid activation where the amino acids are attached to transfer RNA molecules.

On one end of the tRNA molecule is a three-nucleotide anti-code. This will be able to match up with a complementary three-nucleotide code on the mRNA. On the other end, an amino acid is attached that is specific for the tRNA carrying the appropriate three-nucleotide code. An enzyme known as amino acid activating enzyme (or aminoacyl-tRNA synthetases) energizes each amino acid in order to attach it to the opposite end of the tRNA. The specificity of which amino acid goes with which tRNA is determined by the specificity of the synthetases. Once the amino acid has been attached to one end of an adaptor, it needs to be linked into a chain with other amino acids in a specific order to produce a specific protein.

The first tRNA carrying a start codon binds to the free 30S subunit at the peptidyl site. Then the mRNA with the correct start anticodon attaches to the 30S subunit followed by attachment of the larger 50S ribosome subunit. In some prokaryotes, the mRNA binds first to the ribosome followed by the binding of the initiator tRNA to the ribosome. The second tRNA enters the second aminoacyl site. The amino acid carried by the first tRNA is shifted over and bound to the amino acid carried by the second tRNA. The mRNA then shifts to the right and the first tRNA drops off while the second tRNA shifts into the peptidyl site leaving the amino acyl site free to accept another tRNA carrying another amino acid. One by one the triplets are read and the protein chain grows until the final triplet signals "stop" with no adaptor to fit it. The ribosome then separates and releases the mRNA. The protein chain formed then undergoes self assembly as it folds and associates with other protein chains to form enzymes and structural proteins.

Synthesis of Lipids

Fatty acid synthesis is catalyzed with fatty acid synthetase using acetyl-CoA and malonyl-CoA as the substrates and NADPH as the reductant. Malonyl-CoA comes from an ATP-driven carboxylation (addition of CO_2) of acetyl-CoA which comes from glycolysis. Both acetate and malonate are transferred from coenzyme A to the sulfhydryl group of the acyl carrier protein (ACP) that carries the growing fatty acid chain during synthesis. The synthetase adds two carbons at a time to the carboxyl end of the growing fatty acid chain in a two-stage process. First, malonyl-ACP reacts with a fatty acyl-ACP and gives off CO_2 and a fatty acyl-ACP that is now two carbons longer. Glycerol can be esterified to fatty acids to result in phosphatidic acid and then to phospholipids which are essential to membrane function.

Synthesis of Peptidoglycan

Peptidoglycan consists of long polysaccharide chains made of *N*-acetylmuramic acid (NAM) and N-acetylglucosamine (NAG). To the NAM groups, pentapeptide chains are attached. The chains are then attached to each other either directly by the pentapeptide chains (gram negatives) or via a pentaglycine bridge (gram positives). During synthesis of this complex cell wall constituent, two key carriers participate: uridine diphosphate (UDP) and bactoprenol. Bactoprenol is a large 55-carbon alcohol that attaches to the NAM by a pyrophosphate. It moves peptidoglycan subunits back and forth through the hydrophobic membrane.

Eight stages occur in the synthesis of peptidoglycan. UDP derivatives of NAM and NAG are made in the cytoplasm followed by the addition of five amino acids to the UDP-NAM molecule with the expenditure of ATP. This NAM-pentapeptide is then transferred from UDP to the bactoprenol carrier at the inner membrane surface. UDP- NAG then adds NAG to the NAM-pentapeptide to form a peptidoglycan unit. In gram positives, the pentaglycine bridge may also be added at this time. The fifth step involves transporting this unit across the membrane to the outer surface of the membrane by the bactoprenol carrier. The peptidoglycan unit is attached to the growing end of a peptidoglycan chain to lengthen it by one repeating unit, and the carrier, now free, can return to the inside of the membrane to accept another NAM-pentapeptide. Finally, peptide-cross links form between the peptidoglycan chains by transpeptidation.

Peptidoglycan synthesis is one of the main sites of attack by antibiotics. There are several sites of attack available. Penicillin stops the transpeptidation reaction and bacitracin blocks dephosphorylation of bactoprenol pyrophosphate.

Growth

Growth refers to an increase in cellular constituents to result in either an increase in a microorganism's size, population number, or both. In bacterial growth, it usually refers to population increases due to binary fission.

The Growth Curve

When organisms are grown in batch culture, they are incubated within a closed flask with a fixed amount of medium. During growth, the nutrients available are used up and the wastes from metabolism build up. Growth of a bacterial population can be followed by plotting the logarithm of live cell numbers against the incubation time to result in a typical growth curve with four distinct phases: lag, exponential, stationary, death.

When microorganisms are first introduced into fresh media, no immediate increase in cell numbers occurs. This phase is referred to as the lag phase. All cultures go through a lag phase if the medium is different than the one the organisms were transferred from or when the organisms are not in exponential phase. The lag phase may be due to cells being old and depleted of ATP, essential cofactors, or ribosomes which must be synthesized before binary fission can occur. The lag phase is especially long if the inoculum is from an old culture. Sometimes the lag phase can be avoided entirely if it is transferred during log phase into a fresh medium of the same composition.

During exponential phase, the population divides at its maximal rate possible by doubling at regular intervals. Since the population is composed of individuals which are dividing nonsynchronously, the curve rises smoothly. However, the individuals are remarkably uniform during this phase. It is during this phase that a culture should be harvested for use in most studies including preservative challenge tests. Waiting longer (until stationary) allows the cells to become more resistant to the biocide or preservative.

Ultimately, the population reaches the stationary phase where the total number of live microorganisms remains constant due to cell death rate equaling the cell division rate. Alternatively, it may be due to the cells simply ceasing to divide despite remaining metabolically active. Organisms enter the stationary phase due to several reasons: nutrient limitations, oxygen limitation, or the accumulation of toxic waste by products such as acids. These factors may occur singly or in concert.

During death phase the viable cells decrease exponentially as they die. This rate of death may decrease over time due to the extended survival of particularly resistant cells.

Mathematics of Growth

During the exponential phase, the population is dividing at constant intervals and doubling during a specific length of time called the doubling time or generation time. Because the population is doubling every generation, the increase in population is always 2^n where n is the number of generations. To understand the concept of exponential increase, one can use a financial example of a pay scale where one receives a penny the first day of work which doubles thereafter each day. After 30 days, the total pay received is $10,737,418.24. The equation for this investment where one's money is compounded daily at 100% interest for 30 days is the same as that for a bacterium which is "compounded" every 20 minutes at "100% interest." That equation is $N_T = N_O \text{ X } 2^n$, where N_O is the initial population number, N_T is the population number after some time T, and n is the number of generations in time T.

Knowing this rate of growth allows one to appreciate the concern when a lab finds that a single organism exists in a mix tank of 30,000 gallons (75,000 liters). Over the weekend, that one organism dividing every 20 minutes could result in 1.05×10^{65} organisms in the tank or 10^{57} organisms for every milliliter in the 75,000 liter tank. Of course, the culture would have already reached stationary and entered death phase with no more than 10^9 cells per milliliter as a maximum long before such high numbers were reached. In fact, the 10^9 cells per each of the 75,000,000 milliliters in the 30,000 gallon mix tank was reached after 50 generations or just less than 17 hours after the discovery of the one organism in the tank.

Measurement of Cell Numbers

One of the easiest ways to determine the numbers of microorganisms in a sample is to count them directly under a microscope. In order to count them accurately, a Petroff-Hauser counting chamber can be used since they have chambers of known volume. The disadvantages of using microscopic counts is that if the population is dilute, then the count may not be accurate. Operator fatigue can be significant and overcome with the use of image analysis and recording films. Finally, the ability to distinguish live and dead cells is difficult but can be overcome somewhat with viability stains.

Use of particle counters such as a Coulter Counter works for quantifying protozoans, algae, and nonfilamentous yeasts. Bacteria are not countable with it due to interference from particles and filaments. As the organism passes through a small hole, an electrical current is interrupted which sends a signal to be counted.

Perhaps the most used technique is to count microorganisms (bacteria especially) by the plate count. A diluted sample containing the bacteria is spread over a medium of nutrient agar contained in a Petri plate. After incubation, the organisms grow into large collections of several million organisms which can be seen and counted. In this way, an estimate of the numbers in the original sample can be made based on colony forming units. This estimate is based on the fact that bacteria rarely exist as separate cells in nature but as clumps of cells. Thus, when they are spread out on a plate, the units that form a colony are often composed of several organisms stuck together rather than individual organisms. Membrane filters can also be used to trap microbes as they are filtered through the filter. The filter is placed onto an agar surface and the nutrients are absorbed by the filter to provide the organisms a chance to grow into colonies.

Continuous Culture

In batch culture, the wastes are not removed and the nutrients are not replenished. The typical growth curve exists where the exponential phase last only a few generations. In nature, however, many systems are open with nutrients flowing into the system and wastes flowing out of the system. Such systems can be built in the laboratory as continuous cultures. These systems are very useful for studying interactions between microbial species and for mimicking natural lake and stream ecosystems. They are even useful for understanding growth in manufacturing systems which resemble plug and flow continuous cultures.

In continuous cultures, the flow rate can be varied as it flows into a constant volume container such that for every drop of fresh medium that drips into the culture vessel, a drop of spent medium drips out. A microbial population can be maintained in the exponential phase and at a constant biomass for as long as the medium continues to flow into the culture vessel. The rate of dilution is controlled entirely by the flow rate and the volume of the vessel. In such a system, which mimics natural systems as well as manufacturing systems, growth is controlled by the limiting nutrient. The dilution rate controls the rate of nutrient provided and is related to the flow rate in ml/hr (*f*), and the volume of the culture vessel (*V*) by the following relationship:

$$D = f/_{V}$$

The dilution rate has units of hr^{-1}. As the dilution rate increases, the population's growth rate increases to match it until the dilution rate exceeds the growth rate and the organism washes out faster than it can reproduce. At dilution rates that are just below the washout rate, the organisms increase in both cell density and growth rate compared to dilution rates that are lower where cell density is lower and growth rate is also slower. At slow dilution rates, only a limited amount of nutrients are available and nearly all the energy derived from those nutrients is used up quickly just for cell maintenance rather than for growth and reproduction. As the dilution rate is raised and more nutrients are provided, energy becomes available for growth as well as maintenance. Thus there is a certain amount of maintenance energy that must be exceeded before growth can take place.

Environmental Growth Conditions

Often the only way we control microbial growth is through biocides and preservatives. However, we should also consider all the chemical and physical parameters that must be provided to allow for growth to occur. Understanding these parameters may allow us to rely less on biocides and more on creating products that are inherently hostile to microbial growth because of their physical and chemical composition even if no preservatives are added. Thus, we may be able to limit growth in a manufacturing plant with these same parameters and reduce our reliance on chemical biocides.

Solutes and Water Activity

Most microorganisms exist in hypotonic solutions in nature. Thus, water enters the cell through the semi-permeable membrane and would cause it to burst if it were not for the rigid cell wall that maintains the shape and integrity of the organism. Clearly, organisms without cell walls such as *Mycoplasma* require isotonic conditions for growth. Many bacteria actually concentrate the content of their protoplasm in order to keep water flowing into them thus keeping their plasma membrane pressed firmly against the cell wall. Bacteria increase their osmotic concentration by synthesizing compatible solutes such as choline, betaine, amino acid, and potassium ions.

A far more devastating condition for many bacteria is hypertonicity. Anytime the solutes are more concentrated on the outside of the cell than the inside, water leaves the cell and the plasma membrane shrinks away from the wall; this is known as plasmolysis. This effectively constitutes dehydration and may damage the cell, especially the membrane where energy processes occur. The cell then becomes metabolically inactive and so stops growing. Such conditions can be used to control the growth of organisms.

One can actually reduce the amount of water available to the microorganisms by interaction with solute molecules or surfaces. In the former, the water is bound by the solutes in the environment whereas in the latter, the water is adsorbed to surfaces of solids. Thus, one cannot assume that simply because an environment has plenty of water in it that the water is available to the organisms. Expressing quantitatively how much water is available to an organism is done by the relating the vapor pressure of a solution to the vapor pressure of water:

$$A_W = P_{soln} / P_{water}.$$

Another way of thinking about water activity is to realize that it is a measure of the relative humidity of the solution if the solution were placed in a sealed chamber after the chamber came to equilibrium. Water activity is inversely related to the osmotic pressure of a solution. The lower the water activity, the higher the osmotic pressure and the less likely organisms are to grow. Most organisms prefer growing at water activities of 0.98 to 1.0. A few are osmotolerant and can survive water activities as low as 0.6, but these are generally nonpathogenic. Addition of a wide variety of water binding molecules in a formulation can actually tie up the water resulting in a low A_W. Thus, this avenue of preservation should not be overlooked. For example, many antiperspirants and dry make-up powders have extremely low A_W and do not support growth unless water is in the skin debris and is placed into the products during use. As a result, preservative challenge tests are meaningless for these products and should be replaced with water activity measurements and consumer use data. Table 2-2 shows some key water activity points and the organisms inhibited by low water activity and the effects of pH, a controlling factor described further below.

pH and Temperature

pH measures the hydrogen ion activity of an environment on a logarithmic scale from a pH of 0 to 14 where 0 represents the high concentration of 1.0 M hydrogen ions (H^+) and 14 is the low concentration of hydrogen ions of 10^{-14} ions. Microorganisms grow in environments ranging from a pH of 1 to 10. The only microbes of concern for microbiologists involved in keeping consumer products free of microorganisms are the neutrophiles growing at 5.5 to 8.0 since these are also potential pathogens. Unless

TABLE 2-2

WATER ACTIVITY AND THE POTENTIAL FOR GROWTH

Water Activity	pH	Problem Organisms Capable of Growth	Examples of Cosmetic Products
0.98 to 1.00	pH 5 to 9	Most Gram positives and negatives	Shampoos and emulsion products
0.95 to 0.97	pH 5 to 9	Most Gram positives and negatives (*Pseudomonas* begins to be limited)	Liquid make-ups and eye area products
	Below 5.5	Some Gram negatives and most Gram positives (*Pseudomonas* limited)	Some hair conditioners
0.92 to 0.95	Above 5.5	Few Gram negatives and most Gram positives	Some pressed powders
	Below 5.5	Most Gram positives	
0.90 to 0.92	pH 5 to 9	Gram positive Lactobacilli and *Staphylococcus*	Some rouges (non-water based)
0.80 to 0.90	pH 5 to 9	*Staphylococcus,* Molds, Yeasts	Lipsticks (non-water based)
0.70 to 0.80	pH 5 to 9	Molds, Yeasts	Some talcs
0.65 to 0.70	pH 5 to 9	Osmotolerant yeasts	Some antiperspirants
0.60 to 0.70	pH 5 to 9	Osmotolerant and xerophilic molds	
Below 0.60	pH 5 to 9	None	

spoilage organisms are present that grow below 5.5 or above 8.0, these provide minimal concern to the consumer product microbiologist. Table 2-3 shows the pH ranges of microbes and consumer products/environments where the various pH values fall.

Many products do not support the growth of potential pathogens if the pH is below 4.5 or above pH 10. The only reason for adding biocides to these products is to control potential spoilage organisms.

Temperature affects microorganisms since they have no way to control their own temperature and because enzymatic reactions required for growth are affected by temperature in the same way any other chemical reaction is affected: the rate of reaction roughly doubles for every 10°C rise in temperature. At low temperatures, the enzymes operate more slowly and thus growth is also slower. Up to a certain temperature, growth increases until the point is reached where high temperatures are lethal as they damage the cell by denaturing proteinaceous enzymes, destroying

TABLE 2-3

pH RELATED TO THE TYPES OF ENVIRONMENTS AND CONSUMER PRODUCTS AND THE TYPES OF ORGANISMS PRESENT IN THOSE ENVIRONMENTS

pH	Environment/Consumer Product Example	Organism Present
0	Concentrated Acids	A few thiobacilli
1	Gastric Acids; acid thermal springs	A few thiobacilli and thermoacidophiles such as *Sulfolobus*
2	Lemon juice	*Thiobacillis*
3	Vinegar	A few lactobacilli, some *Leuconostoc*
4	Tomatoes, orange juice, hair conditioners	*Lactobacillus* Some staphylococci, yeasts and molds
5	Cheese, bread, skin lotions, and make-ups	Some pseudomonads, *Lactobacillus acidophilus*
6	Beef, chicken, milk, shampoos	*Pseudomonas aeruginosa*
7	Water, blood, shampoos, mascaras	*Staphylococcus aureus, Escherichia coli*
8	Seawater, some shampoos	*Nitrosomonas, Bacillus*
9	Alkaline soils and lakes, some deodorants, antiperspirants, and skin cleansers	*Anabaena*, some bacilli
10	Soap	*Microcystis*, some bacilli
11	Household ammonia	Some bacilli survive
12	Drain openers	Spores survive
13	Bleach	Spores sometimes survive
14	Saturated sodium hydroxide	Spores sometimes survive

transport proteins, and destroying lipid membranes. The key or cardinal temperatures to consider for each species of microbe are the minimum, optimum, and maximum. Temperature ranges for the major groups of organisms of concern are shown in Table 2-4.

Microorganisms can be placed into four categories based on their temperature ranges. Psychrophiles grow at 0°C up to 20°C and are found throughout Arctic and Antarctic regions as well as manmade refrigerated environments. In alpine snowfields, the psychrophilic alga, *Chlamydomanos nivalis*, produces pink snow. Psychrophilic bacteria include members of the genera *Pseudomonas*, *Flavobacterium*, *Achromobacter*, and *Alcaligenes*. Facultative psychrophiles grow optimally at 20 to 30°C and can spoil refrigerated foods. Few, if any, of these organisms present a problem for consumer products since they only act as spoilage organisms and consumer products are rarely at low temperature. Psychrophiles and

TABLE 2-4

TEMPERATURE RANGES FOR KEY MICROORGANISMS FOUND IN CONSUMER PRODUCTS IN COMPARISON WITH KEY EXTREMOPHILES

Microorganism	Minimum Temp. (°C)	Optimum Temp. (°C)	Maximum Temp. (°C)
Micrococcus cryophilus	–4	10	24
Pseudomonas cepacia	4	25 to 30	40
Staphylococcus aureus	6.5	30 to 37	46
Enterococcus spp.	0	37	44
Escherichia coli	10	37	45
Pyrodictium occultum	82	105	110

psychrotrophs are not pathogens, although some produce toxins that, if ingested, may cause problems. Mesophiles can grow from 20 to 45°C although their optimum is between 25 to 40°C. Nearly all human pathogens are mesophiles because we are at 37°C. Mesophiles are the key problem organisms for the consumer product microbiologist to control. The key organisms to control in manufacturing environments that are mesophiles include *Pseudomonas cepacia*, *Pseudomonas maltophilia*, *Pseudomonas aeruginosa*, *Enterobacter cloacae*, *Enterobacter agglomerans*, *Enterobacter gergoviae*, *Escherichia* spp., *Staphylococcus* spp., and some yeasts and molds. Thermophiles grow from 45 to 110°C although the typical range is 55 to 85°C.

Oxygen

Many of the organisms encountered in a manufacturing environment are obligate aerobes, organisms that grow in the presence of atmospheric oxygen (O_2). Oxygen is critical for these organisms since it serves as the terminal electron acceptor for the electron transport chain that generates ATP during aerobic respiration. Other organisms are microaerophiles that require some oxygen but not at levels encountered in the atmosphere. Some organisms are facultative anaerobes; they do not require oxygen but do better when they have it. Aerotolerant anaerobes grow whether or not oxygen is present since they are unaffected by it. Occasionally, but relatively rarely, encountered in a manufacturing environment are obligate anaerobes, organisms that grow in the absence of oxygen. Although strict anaerobes are killed by any exposure to oxygen, they can be recovered from habitats that appear to be aerobic. This is especially true for organisms that associate with facultative anaerobes in communities where the facultative anaerobes are serving as oxygen-scavenging guilds to deplete any available O_2. A summary of these oxygen requirements and typical examples of them are shown in Table 2-5.

TABLE 2-5

OXYGEN REQUIREMENTS AND PRESENCE OF KEY ENZYMES IN SELECTED BACTERIA

Oxygen Tolerance	Superoxide Dismutase	Catalase	Examples
Obligate aerobes	+	+	*Pseudomonas* spp. *Bacillus* spp.
Facultative anaerobes	+	+	*Staphylococcus* spp. *Enterobacter* spp.
Microaerophiles/ Aerotolerant anaerobes	+	–	*Enterococcus* spp.
Obligate anaerobes	–	–	*Clostridium* spp.

One of the key reasons for these differences in oxygen tolerance among organisms is due to the effect of toxic oxygen byproducts of metabolism. Oxygen accepts electrons and then becomes reduced into superoxide radical, hydrogen peroxide, and hydroxyl radical prior to being broken down into water. Organisms that are obligate aerobes need enzymes that protect against these toxic O_2 products. Such organisms have evolved the enzymes superoxide dismutase and catalase as they developed the more efficient mechanisms of energy generation using an oxygen-based metabolism. The relationship between the presence of these enzymes and oxygen tolerance is also shown in Table 2-5.

Diversity

One of the fascinating and somewhat bewildering aspects of the microbial world is its extreme diversity. The microbes, it. seems, have the ability to adapt to nearly any environment given them. As a result, a variety of physiologies have evolved such that some species break down sugar while others use sulfur and iron for energy. Some live at ambient temperatures and others live at either freezing or boiling temperatures. Some need organic compounds for growth, while others fix carbon dioxide. No less amazing is the ability of these organisms to adapt to biocides. The diversity of these organisms is due to the ability of these organisms to evolve so rapidly owing to the fact that generation times are so short and the fecundity of the organisms is so great.

Several factors contribute to the ability of microorganisms to adapt and become tolerant to biocides. The classic tolerance mechanism is based on mutation and plasmid resistance which we discuss below. However, there are the physical mechanisms of growing within a biofilm for protection from the biocide or even growing within a community of several species where some organisms provide protection for others within the community. There is also the classic case of Darwinian selection of some phenotypic trait that contributes to tolerance; a bacterium may even overexpress a specific gene for the production of some trait that confers tolerance against a biocide without changing its genotype. These mechanisms are simple hairy elephant stories: when it gets colder, only those elephants with more hair in the normally distributed curve will survive to reproduce. During exposure to biocides, only those bacteria that have a thicker capsule survive for subsequent reproduction and the population shifts to those that produce plenty of slime.

The Role of Mutation in Bacteria

Mutations are usually detrimental to organisms. Typically, for life to exist with stability, the nucleotide sequence of genes must not be disturbed. However, occasional sequence changes occur which are not detrimental. These generate new variability and, if they by chance confer some selective advantage to the organism, they allow for evolution of the population into a new strain or even a new species that can better survive the new environment. If that new environment is one with a biocide, then a tolerant population results. Mutations occur either spontaneously in the absence of any mutation-inducing agents or as a result of exposure to a mutagen. In the latter case, if the biocide also shows mutagenic properties, then it may actually "direct" the microbe to mutate (however, in a completely random way) and thus evolve resistance in what appears to be an almost Lamarckian mechanism.

Spontaneous mutations may be transition mutations where the wrong base is paired with the template base. For example, the purine base adenine or guanine may pair with the wrong pyrimidines, cytosine or thymine, respectively, due to tautomeric shifts. Although very rare, mutations may also be transversion mutations where a purine is substituted for a pyrimidine or vice versa. Spontaneous mutations may also result from frameshifts caused by the deletion of DNA segments to result in an altered codon reading frame.

Induced mutations are caused by agents that: (1) damage the DNA, (2) alter its chemistry, or (3) interfere with its repair mechanisms. These agents, mutagens, act by using base analogs in place of the base called for, mispairing of the base, or intercalating a base into the DNA. If the damage occurs to the DNA before it can be repaired and a complete DNA replication cycle takes place before the site is repaired, then the mutation can become stable and inheritable if it confers some selective advantage for the organism.

Recombination in Prokaryotes

The ability to recombine genetic information in sexually reproducing organisms is an efficient way to insure that diversity occurs in order that adaptation to changing environments (e.g., evolution) can have a chance. The use of a biocide represents a drastic change in the environment for a microorganism. Fortunately for the bacterium, there are adaptation mechanisms in place to overcome the environmental disturbance caused by the biocide. Many of these adaptation mechanisms are due to mutations as described above. However, these are relatively ineffective in maintaining the diversity needed. Certainly, once the mutation occurs in a single bacterium and it reproduces rapidly, the favorable gene is spread throughout the progeny of the asexually reproducing clone. But the bacteria need a way to transmit the favorable mutation in more efficient ways. Since bacteria do not use sexual reproduction, how do they do this to enhance their diversity? How do they adapt to changing environments? Keeping in mind that the key goal in sexual reproduction is recombination of genetic material, just how do bacteria achieve gene recombination without sex?

The most common way bacteria do this is via general recombination involving a reciprocal exchange between pairs of homologous DNA sequences. It occurs anywhere on the chromosome where a homology exists with another DNA strand where a break and subsequent crossover can occur. Many of the enzymes involved in carrying out this recombination are the same as those involved in DNA repair. In bacterial transformation (detailed below), a nonreciprocal recombination occurs where sections of genetic material is inserted into the chromosome by incorporating a single strand of DNA that forms a stretch of heteroduplex DNA. Although the one strand is homologous, it is inserted into the genome as a complementary strand to the original and not duplicated. A second type of recombination is site-specific recombination typified by the integration of a virus genome into its host cell's genome. The genetic material is not homologous with the bacterial chromosome and in fact it is a mechanism to integrate the genome into the bacterial chromosome using enzymes unique to the virus and its host. A third type of recombination is replicative recombination that is typified by transposons where a gene is replicated and then moved about the chromosome without regard to homology.

Movement of DNA from a donor bacterium to a recipient can occur three ways. Direct transfer between bacteria by physical attachment via a pilus (conjugation), transfer of DNA from an environmental source to the bacterium (transformation), and transport of the DNA from one bacterium to another via bacteriophages (transduction). Once the DNA enters the cell, it may be degraded (host restriction), be integrated into the chromosome especially if homologous, replicate outside the cell's normal genome, or not replicate.

Plasmids and Transposons

Plasmids are small (usually less than 30 genes), circular DNA molecules that exist independently of the main bacterial chromosome. They replicate completely autonomously from the chromosome and are stably inherited. Some bacteria have multiple copies of the same plasmid, presumably allowing the genes to be transcribed more often if a need for the proteins for which they code is great. Bacteria can survive without their plasmids because, for the most part, the plasmid carries information not critical for the bacterium's survival.

The classic examples are plasmids that carry R factors (resistance factors) that code for enzymes to destroy or modify antibiotics. If these resistance genes are within a transposon, bacteria can rapidly develop multiple resistance plasmids. If the plasmids are conjugative, they can spread throughout a population and even between species of bacteria. Some plasmids are virulence plasmids that cause the bacterium to better resist host defenses or produce toxins. Others are metabolic plasmids that carry genes for enzymes to degrade specific substances.

Transposons are pieces of DNA that can move around the genome without regard to homologous sites on the chromosome. They can originate in one location on the chromosome and move to a new location on the same chromosome. The actual mechanism of this process has been studied in detail but we will not cover it here. Transposons produce a variety of important effects. They can insert into a gene to cause a mutation that destroys the expression of certain genes of the host or the loss of those genes. Other times, they stimulate overexpression of the gene if they carry promoters and activate the genes near to where the transposon inserts. Thus, they can turn on or off many genes throughout the chromosome depending on where they insert. In fact, the many colors of Indian corn and the mosaic of colors on the skin of an apple are due to these jumping genes going from one spot to the next on the chromosome and telling it which colors to express for its particular cell. Transposons also carry antibiotic resistance genes and can even cause a plasmid to accept additional resistance genes if the transposon carrying the genes transfers into the plasmid. Because the transposons also jump from plasmids to the chromosome, they can cause drug resistance genes to exchange from the plasmid to the bacterial chromosome.

Conjugation

This mechanism of recombination in bacteria looks like sex but it is not. In conjugation, bacteria exchange portions of DNA via a pilus from one to the next. There are even positive "mating" types that are more capable of transferring a copy of a portion of their genes to another bacterium. Since they are haploid, however, their chromosome number (a single circular chromosome) is never halved and then recombined with another chromosome from the opposite sex as in true sexual reproduction. For a bacterium, there is no opposite sex! Conjugation is an efficient way to transfer genes from one bacterium to another.

Transduction

Bacteria, as all living creatures, have viruses that attack them. Viruses are particles of protein and nucleic acid that use the cell's biochemical factory to replicate more viruses. The nucleic acid injected into the bacterial cell from the virus directs the cell to stop doing cell-stuff and instead to make viruses. Viral replication, however, is rather sloppy. It is a process of self-replication where all the viral parts are built and then they come together on their own to form a viral particle simply due to the parts being chemically attractive to one another. This is almost a case of putting all the watch parts together in a box, shaking it, and finding that the box now has a complete watch inside with all the parts simply coming together on their own; the only difference is that the viral parts have only one way that they can fit together and they have chemical properties that cause the parts to be attractive to each other in very specific ways. Some viruses, however, are very sloppy in this random shuffling together of parts and sometimes incorporate bits and pieces of the bacterial host's own DNA rather than the virus nucleic acid. When this happens and the virus goes on to the next cell, instead of the bacterial host forming new viruses, it gets a piece of the DNA from the previous bacterium. Since the DNA for making the virus is incomplete, no new viruses are made in the bacterium but it now has a new gene from another bacterium. The virus was an unwitting patsy in transferring genes from one bacterium to another. If those genes confer some useful property on the recipient, then it is more likely to survive and reproduce to spread the newly acquired gene throughout the population.

TRANSFORMATION

Transformation is one of the more unusual mechanism of sharing genes that bacteria have. They simply absorb DNA from the environment. This requires a cell to lyse and spill its genes into the environment, and it requires a competent cell to absorb the naked DNA from the environment. Oddly enough, the cells are only competent during the exponential phase of their growth cycle when they are also most susceptible to biocides. The DNA that can be absorbed can be either linear or as a plasmid. The efficiency of absorbing plasmid DNA is greater than absorbing linear DNA because they are not degraded as easily.

Selected Bacteria of Industrial and Medical Importance

The organisms most likely to be encountered in contaminated cosmetic products are those that are likely to be present in an ordinary household. In a study of contaminants of cosmetics, the typical organisms involved are those detailed in Table 2-6.

PSEUDOMONAS

Pseudomonas aeruginosa, *Pseudomonas maltophilia*, *Pseudomonas paucimobilus*, and *Pseudomonas cepacia* are ubiquitous microbes found in a variety of environments. They are also quite versatile in their nutritional capabilities and adaptational abilities. They seem to adapt quickly to any new exposure to toxic chemicals including biocides. The

TABLE 2-6

TYPES OF MICROORGANISMS ISOLATED FROM COSMETICS AFTER USE OR FOUND IN CONTAMINATED COSMETICS

Gram-negative Nonfermentors	Gram-negative Fermentors	Gram-positives	Yeasts	Molds
Acinetobacter spp.	*Citrobacter freundii*	*Bacillus* spp.	*Candida*	*Absidia*
Alcaligenes spp.	*Enterobacter cloacae*	*Staphyloccus aureus*	*Saccharomyces*	*Alternaria*
Pseudomonas cepacia	*E. agglomerans*	*S. epidermidis*	*Torula*	*Aspergillus*
P. putida	*E. aerogenes*	*Enterococcus* spp.	*Zygosaccharomyces*	*Citromyces*
P. fluorescens	*E. gergoviae*	*Micrococcus* spp.		*Cladosporium*
P. paucimobilis	*Klebsiella oxytoca*	*Sarcina* spp.		*Dematium*
P. aeruginosa	*K. pneumoniae*	*Streptococcus* spp.		*Fusarium*
P. maltophilia	*Proteus* spp.	*Propionibacterium*		*Helminthosporium*
	Serratia liquefaciens			*Hormodendrum*
	S. marcescens			*Mucor*
	S. odorifera			*Geotrichum*
	S. rubidaea			*Paecilomyces*
				Penicillium
				Phoma
				Aureobasidium
				Rhizopus
				Thamnidium
				Trichothecium
				Verticillium

pseudomonads are infectors of wounds and burns. They also cause pneumonia in patients on immunosuppressive drugs. In cosmetics, the organism has been implicated in eye infections and loss of sight. When found in a cosmetic manufacturing plant, the origin is usually a water system out of control, biofilms forming in the equipment, poor sanitization frequency or poorly done sanitizations, and dead legs or other sources of stagnant product.

Metabolically, the genus *Pseudomonas* is aerobic, using oxygen (and sometimes nitrate) as a terminal electron acceptor. They are Gram-negative rods. They all use the tricarboxylic acid cycle to oxidize substrates to carbon dioxide. Most of the hexoses are degraded by the Entner-Douderoff pathway rather than by glycolysis. Perhaps their more fascinating metabolic capacity is to degrade complex materials including biocides, various hydrocarbon ring structures, and even chlorinated pesticides with ease.

Serratia

This is an organism that is typically not considered to be a pathogen since infections arising from *Serratia* are rare. It is a member of the family *Enterobacteriaceae* (includes *Escherichia*, *Enterobacter*, and *Proteus*), the largest of the three families in section 5 of Bergey's manual that focuses on the anaerobic Gram-negative rods (the *Vibrionaceae* and *Pasteurellaceae* being the other two groups). Like most of the members of this family, they all degrade sugars by means of the Embden-Meyerhof pathway (glycolysis) to pyruvate and then use the pyruvate as the terminal electron acceptor to yield a variety of different end products in a fermentation process. In *Serratia*, pyruvate is further reduced to butanediol, ethanol, and carbon dioxide. *Serratia* has been found contaminating disinfectants and surfactants.

When found in a cosmetic manufacturing plant, the sources to consider first are stagnant product in the system due to dead legs or a breakdown in sanitization frequency or technique.

Escherichia

This organism, like *Serratia*, belongs to the *Enterobacteriaceae*. It carries out a mixed acid fermentation pathway to produce lactate, acetate, succinate, and either formate or hydrogen, carbon dioxide gas, and ethanol. *E. coli* is perhaps the best- studied bacterium and experimental organism of microbiologists. It is a major inhabitant of the human gut and, as such, is a presumptive positive for the presence of fecal contamination in water. Some strains cause gastroenteritis or urinary tract infections. It can grow in the gut, producing enterotoxins that cause the hypersecretion of chloride and water in the small intestine; this is colloquially referred to as "traveler's diarrhea."

Typically, this organism can be found in water systems that are older or ones with considerable biofilm (corrosion) in the pipes. It is also a naturally occurring organism and, despite its use as an indicator of fecal contamination, such a presence in older water distribution systems does not necessarily indicate the presence of feces in the water.

Enterobacter

Enterobacter, like *Serratia* and *Klebsiella*, produces butanediol, ethanol, and carbon dioxide. It can be isolated from contaminated surfactants and particularly quaternary-containing conditioners. It is also a typical contaminant in households and can grow in poorly preserved products. Some of the key species contaminating cosmetics include *Enterobacter agglomerans*, *Enterobacter gergoviae*, and *Enterobacter cloacae*. They are found in soil and often invade plant tissues causing a variety of necroses. They are not generally considered to be human pathogens unless directly introduced into the blood stream. It is a Gram-negative fermentative rod.

Klebsiella

These organisms are very widespread in the environment. They are human pathogens, while some of the species are commensals. They are found in soil and water and are plant pathogens. *Klebsiella pneumoniae* causes a severe fulminating pneumonia in people who are debilitated either physically or due to alcohol abuse. Unlike many of the enterobacters, *Klebsiella* is nonmotile. *Klebsiella* is found routinely in households and can contaminate cosmetics during consumer use. It is a Gram-negative rod in the *Enterobacteriaceae* family.

Proteus

These are motile, produce hydrogen sulfide gas, and often exhibit a type of motion that is called swarming. They decompose urea to produce ammonia and carbon dioxide. *Proteus* is usually found in the intestinal tract of man and animals, and is found in areas contaminated with feces. *Proteus* is typically associated with urinary tract infections; however, they can cause pyogenic infections in other parts of the body if accidentally introduced (e.g., wound infections). It is a Gram-negative rod. Its presence in a cosmetic would indicate either contamination of a raw material like surfactant, or the use of water with a high level of contamination.

Staphylococci

These Gram-positive organisms are cocci which are nonmotile and can grow aerobically or anaerobically. Some of the species (e.g., *S. aureus*) cause boils, are involved in impetigo, cause conjunctivitis, and cause food poisoning. A common manifestation of infection is the production of pus. When cultured on blood agar, *S. aureus* produces a clear zone of hemolysis known as β-hemolysis. The reason for this hemolysis is due to the production of a toxin that lyses red blood cells. The role of the hemolysin in disease is not entirely known. The staphylococci also produce a variety of enzymes such as hyaluronidase, proteinases, lipases, coagulase, and penicillinase. The production of coagulase can be used to confirm in the laboratory that the organism is *S. aureus*, a potential pathogen. When *S. aureus* produces coagulase, it clots the serum. Performing this test in the lab is relatively simple with the addition of a culture to human or rabbit plasma in a tube; after a brief incubation, the coagulase-positive *S. aureus* clots the plasma due to the formation of a fibrin clot. Its presence in a cosmetic indicates human contamination.

Streptococci

The streptococci are Gram-positive cocci that appear in chains, sometimes in pairs. It can grow either aerobically or anaerobically. Some produce an alpha hemolysis (green zone) on blood agar plates; others produce a beta hemolysis (clearing of the blood around the colony,) and others produce a gamma hemolysis (no effect). The beta hemolytic strep include *Streptococcus pyogenes.* It is a human pathogen causing strep throat. It can also be a dangerous infective agent in wounds and in blood poisoning after childbirth (puerperal sepsis). Its natural reservoir is in the throats of carriers, usually children. At one time, it was a scourge progressing to scarlet fever (producing erythrogenic toxin) and rheumatic fever because of the lack of antibiotics to control it. Now, it continues to break out occasionally and sporadically.

Streptococcus spp. in cosmetic products is rare. Its presence in a cosmetic would be a result of the lack of good sanitary practices by employees who place their hands into the product or the container.

Bacillus

The genus *Bacillus* are widely distributed saprobes throughout the earth's habitats. They are aerobic and catalase positive. Living primarily in the soil, they are varied nutritionally, but are not fastidious. They are also a common source of antibiotics. They are spore forming organisms that continuously disperse spores into the water and onto plants. In the cosmetics industry, some of the more common raw materials to be contaminated with *Bacillus* spores include aloe vera and a variety of thixotropic agents such as quaternized clays. Pasteurization of aloe vera gel will not eliminate *Bacillus* since the spores are not susceptible. Instead, tyndallization is required where the heating process is repeated several times on three successive days to allow the spores to germinate and become susceptible to the heating processes. Alternatively, irradiation can be used particularly for the clays since they are not harmed by the process.

Very few of the *Bacillus* spp. are pathogenic except *Bacillus anthracis* and *Bacillus cereus. B. anthracis* can cause anthrax, which is a cutaneous disease caused by spores which enter the skin through small cuts and abrasions. It is invasive due to the production of virulence factors that include a polysaccharide capsule and exotoxins that produce edema and cell death. The initial disease presents as a papule that becomes increasingly necrotic and then ruptures to form a painless, black scab called an eschar. *B. anthracis* can also cause pulmonary anthrax from inhaling airborne spores. The onset of this form of anthrax is rapidly fatal within a few hours after septicemia occurs. The toxins produced cause capillary thrombosis and cardiovascular shock. At one time, when natural animal hair was used, shave brushes were significant sources of *B. anthracis*. The transmission by this route was ideal. The spores on the bristles would be placed directly into the skin while the person shaved and nicked himself.

Bacillus cereus is a common airborne and dust-borne contaminant that grows well in cooked foods particularly rice, potatoes, and meats. The organism itself is not pathogenic, but it produces enterotoxins that cause nausea, vomiting, abdominal cramps, and diarrhea which resolve within 24 hours of eating the food contaminated with the toxin. Although not a common organism of concern to cosmetic microbiologists, all species of *Bacillus* are undesirable because they are indicators of filth in the process (e.g., soil) since *Bacillus* spores have a soil origin. In this day of an immuno-compromised population, we also must continue to be aware of the problems that even so-called harmless organisms can offer.

Clostridia

Clostridium is also a Gram-positive, spore-forming rod. It is also widely distributed in soils. Unlike *Bacillus*, it is anaerobic and catalase negative. Many of the members are saprobes growing in sewage, on vegetation, and even as commensals on humans. They are not ordinarily communicable but, instead, infect when they are introduced into the body from puncture wounds or other skin injuries where soil containing spores is introduced into the wound. They can also produce exotoxins that cause an intoxication when the material into which the toxin was produced is ingested.

Wound infections, primarily from *Clostridium perfringens* as well as *Clostridium novyi*, and *Clostridium septicum*, give rise to gas gangrene. Since the *Clostridia* are not highly invasive, they usually gain access to the body via dead or damaged tissues that provide an ideal anaerobic and growth factor-enriched environment. Rapid vegetative growth occurs, and exotoxin plus a variety of enzymes are produced. The toxins released cause extensive tissue destruction. As the organism ferments the carbohydrates in the tissues, gas is formed. Thus, the organisms can easily weaken the tissues with enzymes and toxins and then tear them apart with the production of gas thus enabling the organism to slowly continue its way into the infected patient.

Clostridium tetani causes tetanus, a neuromuscular disease involving the production of a toxin by the organism once it has gained access into the body. It is poorly invasive, requiring not only a wound to gain access but that the wound also is necrotic and poorly supplied with blood since it is a strict anaerobe. Although a localized infection, the toxin that is produced can spread to nearby motor nerves and then travel to the ventral horns of the spinal cord where it binds to the target sites on the spinal neurons responsible for inhibiting skeletal muscle contraction by inhibiting the release of neurotransmitter. Thus, the muscle contracts uncontrollably.

Food intoxication can be caused by *C. perfringens* and *Clostridium botulinum*. However, this route of infection as a result of cosmetic use is low unless the product somehow permits growth and the organisms produce toxin in an oral-care product that is accidentally swallowed (e.g., toothpaste for toddlers). Infant botulism can be caused by ingestion of spores by children less than six months of age where the gut environment cannot kill the spores as it does in adults and older children. The spores survive and germinate and then produce the toxin resulting in "floppy baby syndrome."

Of the *Clostridia*, the main species of possible concern are *C. perfringens* and *C. tetani* since they, like the bacilli, represent a potential source of filth in the product (from soil) and, the product they enter would be deemed adulterated. They would also be considered a potential threat to the consumer's health if the product contaminated by these microbes might be used in such a way that they could infect the customer. A classic example would be products associated with shaving and applied to skin (e.g., shave creams, after shaves and colognes, lotions, antiperspirants, and deodorants).

THE MOLDS AND YEASTS

Physiology and Biochemistry

In a few products where water is limiting and pH is low, the fungi are typical contaminants. Many of these products are lotions and creams. Fungi are eukaryotic: they contain membrane-surrounded organelles within them, particularly a membrane-bound nucleus within which is the DNA comprising the chromosome(s). Other organelles that are membrane enclosed include the mitochondria (which once were bacteria according to the endosymbiont hypothesis), the Golgi apparatus, endoplasmic reticulum, lysosomes, and nucleolus (where ribosomes and ribosomal RNA are made). The term organelle is used to point out the parallel between organs in an animal and the structures within a cell where each performs a very specific function. The extensive membrane systems in eukaryotes are needed because of their large volume and their need for regulation and transport as a result of that large volume.

Cytoplasmic Matrix, Microfilaments, Microtubules

Within the eukaryotic cell is also a homogeneous, rather bland structure known as the cytoplasmic matrix. It is actually the most important and complex structures of the cell since it is here that the organelles exist and where a majority of the important biochemical processes take place. The cytoplasm is composed of 70 to 85% water which is both free and bound to surface of proteins. The pH is about 6.8 to 7.1, except in the digestive vacuoles and lysosomes where it can be as low as 3 to 4.

Within the cytoplasm are microfilaments of about 4 to 7 nm in diameter that provide a structure to aid in cell motion and shape. These filaments have a network or parallel arrangement and have the same basic structure as actin found in our own muscle protein. The other filamentous organelle is a thin cylinder about 25 nm in diameter called a microtubule. They help in shape and cell movement along with the microfilaments, but their major role is in intracellular transport of substances throughout the complex cytoplasm.

Organelles

Endoplasmic Reticulum (ER)

The ER is an irregular network of branching and fusing membranous tubules around 40 to 70 nm in diameter with flattened sacs called cisternae interspersed along the way. A large part of the ER's job is synthesizing protein from the ribosomes located along the surface of the ER. Since this ER is covered with ribosomes, it is referred to as rough endoplasmic reticulum. The other type of ER is smooth ER which lacks ribosomes and is perhaps involved in lipid synthesis. The ER transports proteins, lipids, and other materials throughout the cell. It is also the site of cell membrane synthesis.

Golgi Apparatus

This organelle is made of flattened sacs, called cisternae, that are stacked upon one another. The origin of the Golgi is from the endoplasmic reticulum. The sac that forms from the ER and which faces it is called the cis side; the opposite and already formed maturing face is called the trans side. The membranes composing this organelle are without ribosomes. There may be about 8 or more cisternae in a sac with each one about 15 to 20 nm thick and separated from each other by 20 to 30 nm. At the edges of the cisternae are tubules and vesicles. The main function of the Golgi is to package materials and prepare them for secretion. Apparently, material is transported from the ER to the cis side of the cisternae and then transported to the trans side and on the next cisternae by vesicles that bud off the edges and move to the next sac. Most of the proteins that enter from the ER into the Golgi are glycoproteins which are modified in the Golgi by adding specific groups and sending them on their way to the proper location for use.

Lysosomes

Lysosomes are synthesized by the endoplasmic reticulum and the Golgi apparatus. The digestive enzymes they contain are manufactured by the rough endoplasmic reticulum and then packaged by the Golgi. Lysosomes are spherical single membrane-bound particles about 50 to 500 nm in diameter. They contain all the enzymes (hydrolases) needed to aigest the macromolecules that a cell thrives on. Oddly enough, the internal pH of the lysosome is acidic (3.5 to 5.0) since the unit membrane around the lysosome pumps hydrogen ions into its interior for the proper functioning of the hydrolases.

The key beneficiaries of lysosomes are those cells that obtain their nutrients via endocytosis, where the cell engulfs materials by enclosing them in vacuoles surrounded by a "pinched off" portion of the cell membrane. These vacuoles (or vesicles if they are particularly small) are created by phagocytosis (engulfment of particles) or pinocytosis (ingestion of liquid and the solute molecules). These are called phagosomes or pinosomes; collectively they are called endosomes. The membranes of the lysosomes fuse with those of the endosomes to become food vacuoles as the digestive enzymes of the lysosome mix into the materials carried within the endosome. Once digestion takes place, the nutrients diffuse into the cytoplasm leaving behind a residual body with the indigestible material.

Ribosomes

The basic structure of the eukaryotic ribosomes is similar to that of prokaryotes, except that the subunits are larger in the eukaryotes. They are composed of a large subunit of size 60S and a smaller unit of size 40S for a total size of 80S where S refers to Svedbergs, a measure of how quickly a particle sediments in a centrifuged gradient. Their primary role is translation of the messenger RNA (mRNA) that had been transcribed from the gene within the nucleus of the eukaryotic organism. This translation process is protein synthesis. They are, themselves, manufactured within the nucleolus of the cell.

Within the cytoplasm of a cell, the ribosomes appear as tiny particles that give the cytoplasm a stippled look. However, they are also intimately associated with the rough endoplasmic reticulum (ER), studding its membranes and thus fancifully being considered to look like a biker's black leather jacket with silver studs.

Mitochondria

The mitochondria are currently thought by some scientists to be derived from endosymbionts that were once prokaryotic invaders of large progenitor cells. A symbiosis developed between the large cells which could not manage energy distribution very well; these symbiotic invaders could supply the cell with energy in the form of ATP. There is considerable support for this theory that, during the early formation of life, mitochondria were simply bacterial invaders turned good. Mitochondria have a circular genome-like bacteria; they replicate independently of the cell; and they even have prokaryotic-like 70s ribosomes.

Nucleus

The nucleus of eukaryotic cells, as opposed to the nucleoid of prokaryotic cells, is bound by a membrane called the nuclear envelope. This double-layered membrane is very porous to allow macromolecules easy access in and out. The nucleus contains several linear chromosomes compared to the single circular chromosome of the prokaryotes. It also contains the nucleoplasm and the nucleolus that manufactures RNA for synthesizing ribosomes. The nucleoplasm is a term to describe all enzymes and proteins in the nucleus that are involved in replication of the genome. The key item within the nucleus is the chromatin that makes up the chromosomes, which are composed of DNA and proteins called histones. These are not visible until the cell undergoes mitosis and the chromosomes are duplicated and condensed by forming coils and supercoils around the histones just before being separated into the daughter cells.

External Cell Organelles

Motility and protection are useful characteristics for any organism. Eukaryotic cells enjoy the benefits of motility by means of flagella or cilia. They enjoy the benefits of protection from the external environment due to the glycocalyx.

Cilia and Flagella

The eukaryotic flagellum is composed of microtubules arranged in what is referred to as a 9+2 arrangement. This describes the nine pairs of hollow tubules surrounding a single pair of tubules in the center. The whole arrangement is surrounded by an extension of the cell membrane. The movement can either be a back and forth, whip-like motion or a twirling motion, all as the result of the tubules sliding past each other in an almost muscle like fashion, and all coordinated by the cell membrane. In contrast, bacterial flagella are simple, consisting of a protein strand and undergoing a spinning motion due to a basal body connected into the cell wall. Cilia are composed of the same basic architecture as the flagella, but they are shorter. Usually they cover a large portion of the cell and beat in a regular fashion

External Cell Coverings

Many eukaryotic cells have a glycocalyx as an external boundary layer beyond the membrane (in protozoa and a very few algae) or the cell wall (in most algae and fungi) that comes in direct contact with the environment much like the glycocalyx of prokaryotes. It is made of polysaccharides that compose a slime layer to protect or to provide for attachment to surfaces, even to allow a mechanism for communication between other cells.

The cell walls of fungi and algae provide shape, support, and protection. Fungal cells are composed of a thick inner layer of chitin or cellulose fibers and a thin outer layer of mixed glycans followed by the glycocalyx. The cytoplasmic membrane is the typical bilayer of lipids into which proteins are free to move. The key difference between eukaryotes and prokaryotes is the presence of sterols in the membrane in addition to phospholipids. The sterols add some stability to the membrane by making it less flexible.

Growth and Reproduction

Reproductive strategies in the eukaryotes are more complex than the simple process of binary fission in prokaryotes. Fungi can propagate by simple growth of the hyphae or fragmentation of the hyphae. They can also produce spores. If the spores are a result of mitotic division of the parent cell, they are asexual spores; if they are formed by two parental haploid nuclei fusing and then undergoing meiosis, they are sexual.

Algae reproduce asexually through fragmentation, binary fission, mitosis, and even by producing motile spores. They can also have sexual reproduction like fungi. Protozoans reproduce asexually by mitosis, although a few can use a sexual reproduction referred to as syngamy where two haploid gametes unite to form a diploid zygote which undergoes meiotic division to produce a number of haploid trophozoites (the motile feeding stage). Protozoans can also undergo encystment where a resistant stage much like a bacterial spore is formed. Like the bacterial spore, the cyst is not a means of reproduction. Instead, it is a dormant stage of the organism

Asexual Reproduction in Fungi

The key to identifying fungi is to observe their reproductive structures, the spores. In contrast to the bacteria and some protozoans, a fungal spore is a mechanism of reproduction. There are two types of asexual spores: (1) sporangiospores formed by successive cleavages within a saclike head called a sporangium attached to a stalk and (2) conidiospores which are free and not enclosed by a sac but still develop by segmentation of the vegetative hyphae. Conidiospores come in a variety of forms:

1. Arthrospores (rectangularly shaped; septate hypha fragment at the cross walls)
2. Chlamydospore (spherically shaped; a hyphal cell thickens, fragments, and is released)
3. Blastospore (produced by budding from a parent cell which can be either a yeast cell or another conidium)
4. Phialospore (produced from the budding of a spore bearing structure called a phialide that looks like a vase with flowers which is on top of a conidiophore)
5. Porospore (grows out through small pores in the spore-bearing cell)

6. Micro and macroconidia (small and large conidia that are formed by the same fungus depending on growth conditions).

Sexual Reproduction in Fungi

Sexual reproduction gives an organism the opportunity for increased gene sharing and thus increased diversity. With increased diversity comes increased adaptability and thus increased survivability. The big difference in fungal sex as opposed to most other eukaryotic organisms is that the fungi are already haploid and thus ready for fusing their nuclei to form a temporary diploid state. The most common sexual spores formed are zygospores, ascospores, and basidiospores.

Zygospores are formed when hyphae of two strains fuse to create a diploid zygote which then swells, gets coated by a spiny covering, and is released. When it germinates, it forms a sporangium which then produces haploid sporangiospores with genetic material derived from both parents due to crossing over events while in the diploid state. Ascospores and basidiospores follow the same initial steps of haploid strains fusing as above. However, in ascospores a terminal diploid cell forms at the end of a single ascogenenous hypha that formed as a result of the two parental hyphae (ascogonium and antheridium) fusing. The diploid cell at the tip of this hypha will then undergo meiosis to reduce the genome number down to the haploid state. The haploid cells can then undergo mitosis to produce additional spores, all contained within a sac. In basidiospores, a basidium, or pedestal, is formed upon which the spores are formed. The same processes described previously take place, but now the terminal diploid cell forms the basidium which then produces four haploid cells through meiosis that are borne at the surface of the basidium. This is characteristic of mushroom reproduction where the basidia are along the gills of the fleshy part of the mushroom which we consume. Remember that the next time you are eating mushrooms. Effectively, you are consuming an organism's sexual organs and its progeny at the same time!

Diversity: Fungi of Industrial and Medical Importance

At one time, the fungi did not occupy their own kingdom, the Myceteae. They were classified along with the algae and bacteria as Plantae and later on they were classified with the motile alga and protozoa as Protista. They are unique enough that they eventually were placed into their own kingdom, at least by Whittaker's approach. One still sees the confusion that this indecision led to when trying to explain why some of the slime molds are not protists. The fungi, superficially, can be subdivided into macroscopic varieties such as the mushrooms and puffballs; there are also the microscopic molds and yeasts. Most fungi are unicellular or colonial. A few exhibit actual cellular specialization that approaches tissue differentiation. Some fungi exhibit both a cellular stage and a mycelial stage where they form hyphae; these are called dimorphic. The dimorphic fungi are most often associated with pathogenic fungi, changing from the mycelial form at low ambient temperature to the unicellular form as they infect an animal at higher temperature.

Most fungi are not pathogenic; they are saprobes and free-living without the need for a host. Thus, they are rarely a problem for humans who have a high resistance to them. A few are frank pathogens which can attack healthy people but most are opportunists requiring the right combination of accidental exposure and a weakened immunity for infection to begin. As more and more humans are kept alive by artificial means, or who contract AIDS, or who have cancer or uncontrolled diabetes, the number of immunocompromised people capable of contracting fungal infections will increase. Thus, the cosmetic manufacturer needs to be aware of the use patterns of its products in these populations and plan accordingly. Alflatoxins produced by *Aspergillus* are also a concern for products which are ingested, especially foods whose raw material can support the growth of this mold (e.g., peanut butter).

The fungi comprise a very diverse kingdom subdivided into two subkingdoms, the Myxomycota and the Eumycota. The Myxomycota, or slime molds, play little direct role in cosmetic microbiology. They do, however, play a central role in decay processes of wood, chitin, lignin, and other cellulosic products in nature. Thus, wherever these are used as raw materials in cosmetics, one should be aware that they could serve as food sources for the slime molds given the right environment. Typically, they are found on moist forest floors and appear, as their colorful common names suggest (witches' butter, rutting deer seed), as macroscopically visible (10 to 50 cm) slimy organisms oozing along a rotting log. The Eumycota, or true fungi, include the yeasts and molds we most often associate with human disease and, therefore, play a more direct role in cosmetic microbiology.

The Eumycota are divided into four phyla: Zygomycota, Ascomycota, Basidiomycota, and Deuteromycota. The Zygomycota produce zygospores and sporangiospores or conidia. They include *Rhizopus* and *Mucor* and a variety of others. The Ascomycota produce ascospores sexually and condia when in the asexual mode. There are a variety of pathogens in this phylum including *Microsporum*, a causative agent of ringworm. Also in this group are *Penicillium*, a common source of antibiotic and *Saccharomyces*, a yeast used in beer and bread making. The last of the phyla which produce sexual spores is the Basidiomycota which produce basidiospores sexually and conidia asexually. This phylum includes the mushrooms, bracket fungi one sees one many trees in the woods, and the rusts and smuts that infect plants. The key human pathogen in this phylum is *Cryptococcus neoformans* that causes an invasive systemic infection.

It is interesting to note that *Penicillium*, a cosmetic product contaminant, was once placed into the Deuteromycota as are still some of the species of *Microsporum*; once a sexual stage is observed, the mold is moved from the Deuteromycota to one of the three groups above. The

Deuteromycota are those fungi where a sexual stage has not yet been observed. They produce a variety of conidia, some are dimorphic, and a few are parasitic. The human pathogens include *Candida albicans.* A mildew fungus, *Cladosporium,* is in the Deuteromycota and is capable of infecting cosmetic products kept in the bathroom. *Absidia, Rhizopus,* and *Mucor* can cause mucormycosis and is usually caused by inhalation of dust and soil contaminated by these molds. They are found as contaminants in some products. Several other fungi can cause health problems and contaminate cosmetics. Clearly, antifungal properties of preservatives are important for their control.

Few people can relate to the microbial world because it is so alien and foreign to one's everyday experience. Hopefully, this chapter makes clearer what appears to be a very obscure world: the microbial universe.

SUMMARY

This chapter was intended to give a brief overview of the microbial world in a simple, perhaps even folksey, way so that experts, as well as novices, would gain insight and education. For more extensive references, see the material listed in the references below.[2–8]

REFERENCES

1. Darwin, C. 1859. *On the Origin of Species by Means of Natural Selection or the Preservation of Favoured Races in the Struggle for Life.* John Murray, Albemarle Street, London.
2. Alcamo, I. E. 1997. *Fundamentals of Microbiology.* 5th ed. Benjamin/Cummings Publishing Co., Menlo Park, CA.
3. Madigan, M. T., J. M. Martinko, and J. Parker. 1997. *Brock Biology of Microorganisms.* 8th ed. Prentice-Hall, Inc., Upper Saddle River, NJ.
4. Perry, J. J. and J. T. Staley. 1997. *Microbiology Dynamics and Diversity.* Saunders College Publishing, Orlando, FL.
5. Prescott, L. M., J. P. Harley, and D. A. Klein. 1996. *Microbiology.* 3rd ed. W. C. Brown Publishers, Dubuque, IA.
6. Talaro, K. and A. Talaro. 1996. *Foundations in Microbiology.* 2nd ed. W. C. Brown Publishers, Dubuque, IA.
7. Volk, W. A. and J. C. Brown. 1997. *Basic Microbiology.* 8th ed. Benjamin/Cummings Publishing Co., Menlo Park, CA.
8. Wistreich, G. A. and M. D. Lechtman. 1988. *Microbiology.* 5th ed. MacMillan Publishing Co., New York.

3 MICROBIAL ENVIRONMENT OF THE MANUFACTURING PLANT

Richard Mulhall, Edward Schmidt, and Daniel K. Brannan

CONTENTS

0-8493-3713-5/97/$0.00+$.50

INTRODUCTION

Control of the microbial population within the plant environment is the goal of the plant sanitation program. Good plant sanitation is essential to protect raw materials, the manufacturing process, and the finished product. There are at least eight critical factors that affect the microbial ecology of the plant environment. The three working factors are the cleaning and sanitation practices, personal hygiene of the staff, and raw material handling. The five engineering factors include good sanitary design of the equipment, plant, warehouse, water system, and air system. Current Good Manufacturing Practices (cGMPs) address the various aspects of plant sanitation. Establishment of sanitation standards should include each area and task-related function in the plant.

We discuss these factors and their significance in this chapter. Though we discuss them separately, all the factors relate to each other. Each functions as a part of the complete sanitation and quality assurance program. Inattention to any of these factors will alter the microbial ecology of the plant environment. When this occurs, the material or product becomes susceptible to contamination.

CLEANING AND SANITIZATION BASICS

Each company should set up good housekeeping, cleaning, and sanitization programs for the entire plant environment that meet the specific needs in each area. Good manufacturing practices depend on the cleaning and sanitization practices used and their frequencies of performance within the plant. The specific procedures and schedules for the cleaning and sanitization of equipment and the physical plant should be written.

Several factors determine the effectiveness of cleaning and sanitization. Three of these factors are (1) use of well-designed equipment, (2) use of good cleaning and sanitizing products, and (3) use of validated processes. Finally, management must provide training and supervision of employees. A plant must include timely cleaning and sanitization schedules in its quality assurance program. When it does, there will be insignificant levels of microorganisms in the manufacturing equipment and the plant facilities.

Cleaning

Cleaning is the physical removal of product or ingredient residues. Cleaning removes grease, dust, and contaminants from the surfaces of manufacturing equipment and the environmental surfaces in the building. Physical removal requires the application of energy in some form such as scrubbing, spraying, or turbulent flow. Most cosmetic manufacturers use cleaning agents to aid physical removal. These agents break down soil to let both visible and invisible foreign matter rinse away. An ideal cleaning agent should be readily soluble and should provide good penetration and emulsifying action. It should be compatible with equipment, noncorrosive, and have good wetting and rinsing properties.

Detergent formulations are ideal to use for most cleaning activities. Technically, a detergent is any cleaning agent. However, in popular usage, detergents are washing and cleaning agents with a composition other than a metal salt of an acid derived from fat (a soap). They have much the same mechanism as soap. All detergents have the basic properties of penetration, wetting, dispersion, and emulsifying action.

Detergents may contain surfactants, builders, agents for penetrating, wetting, deflocculating, foaming, emulsifying, sequestering or chelating, and soil dispersants. In addition, these surfactants may be ampholytic, anionic, cationic, or nonionic. Choosing which detergent to be used is based on the intended use and the type of soil to be removed.

Detergent Properties/Ingredients

In describing detergents and their properties, a vocabulary unique to the industry has evolved. The following provides definitions or explanations for a variety of common terms.

Detergency refers to the ability to clean (e.g., soil or other unwanted material from a surface). The detergent does this by a combination of processes including the lowering of surface and interfacial tension, solubilizing or emulsifying, inactivating water hardness, and neutralizing acid soil.

Builders are materials that upgrade or protect the detergency of a surfactant. Builders have several functions. These include inactivation of water hardness, increasing alkalinity to aid cleaning, providing a buffer, suspending soil, and emulsifying oily or greasy soils. Some builders help to **deflocculate** or break up a solid mass into smaller particles and to disperse them through a liquid medium.

Dispersing agents are chemicals that increase the stability of particles in a liquid. The dispersing agent helps remove soil particles by keeping them in a dispersed or suspended state during equipment rinsing. The mechanism of dispersal may be by the process of **emulsification**. This process disperses or suspends fine particles or globules of one or more liquids in another liquid. Similar terms for dispersing agents are **soil suspending agent**s or **inhibitors of soil redeposition**. These are detergent ingredients that keep the soil suspended and dispersed in the cleaning solution. They reduce redeposition of the soil on surfaces by detaching soil globules from a surface and dispersing them through the cleaning solution. Surfactants are the principle emulsifying agents.

Foaming agents are chemical agents that increase the foaming or sudsing characteristics of a cleaning agent. **Penetration** refers to the characteristic that permits the cleaning solution (water) to get under the soil and loosen it from the surface. It also helps the solution to work its way through the soil.

Chelating agents are organic sequestering agents that inactivate water hardness and other metallic ions in water. A commonly used chelating agent is EDTA (ethylene diamine tetraacetic acid and its salts). **Sequestrants** are chelating compounds in aqueous solution that combine with a metallic ion to form a water-soluble combination. In this combination, the ion is inactive. Sequestrants soften water without the precipitation associated with other softening methods like lime softening. For example, complex phosphates are sequestrants that inactivate divalent metal ions such as calcium, magnesium, iron, and manganese without precipitation.

Surfactants (surface active agents) are organic chemicals. We add them to a liquid to change the surface active properties of that liquid. Surfactants and soaps perform the important function of lowering the surface tension of water. Surfactants help remove fatty and particulate soils. They also keep them emulsified, suspended, and dispersed to prevent settling at the surface.

Surfactants are also **wetting agents**. These increase the ability and speed with which liquid displaces air from a solid surface. This property improves the process of wetting the surface. They lower the surface and interfacial tension. This enables the cleaning solution to more quickly wet the surface. In addition to the chemical action of surfactants, one can additionally employ mechanical action to more readily remove soils.

Types of Surfactants

Amphoteric (ampholytic) surfactants may be either anionic or cationic, depending on pH. They are useful because of their wide compatibility with builders, acids, and alkalis.

Anionic surfactants are those whose properties depend in part on the negatively charged ion of the molecule. This property gives us the name anionic. The detergent industry uses a wide range of anionic surfactants that are highly sudsing. Excess foaming is undesirable for surface cleaning. It leaves a residue from excess foam. This residue produces a tacky surface that presents a high resoiling problem.

Cationic surfactants have a positively charged ionic group. Quaternary ammonium compounds are the most widely used cationic surfactants. Their uses include sanitizers and disinfectants, fabric softeners, and static electricity dissipaters. They are not typically cleaning agents by themselves.

Nonionic surfactants contain neither positively nor negatively charged functional groups. They are particularly effective in removing oily soil and many are low sudsing. They do not ionize in water as do anionic and cationic surfactants.

The surfactant industry also has a unique vocabulary for cleaning. We will describe these terms as well. A **general purpose cleaner** contains sodium hydroxide or sodium carbonate for alkalinity, and a sequestering agent. Some cleaners also include a low-foaming wetting agent, and a silicate to inhibit corrosion. These cleaners are different from **clean-in-place (CIP) cleaners** that have the same formula as above, but with a nonfoaming wetting agent. A **manual washing cleaner** will have less alkalinity but will have a high-foaming agent. An **acid cleaner** has an organic or mineral acid to remove hard water and mineral deposits. It may have a heterocyclic nitrogen compound to inhibit corrosion, and a wetting agent for penetration. **Alkaline cleaners** are general purpose cleaners but with very high levels of sodium hydroxide or sodium carbonate. They are ideal for particularly hard cleaning jobs.

Plant personnel should limit the use of cleaning agents in the plant to those necessary for accomplishing the assigned tasks. There is no single all-purpose cleaner, but you should avoid duplication of products wherever possible.

Sanitization

Sanitization is the adequate treatment of surfaces by a process that is effective in destroying vegetative cells of pathogenic bacteria. It also should reduce spoilage microorganisms to insignificant levels. Chemical residuals from such treatment should not adversely affect the product and should be safe for the consumer.

One should precede use of a sanitizer by a cleaning procedure. Alternatively, one can use a combination disinfectant-detergent or detergent-sanitizer in a single-step procedure. One usually reserves this for environmental surfaces. Typically, combinations either have compromised detergency or disinfectant activity compared to using each agent in two separate steps.

Sanitizers are EPA-registered chemicals that reduce viable microbial contaminants on surfaces to safe levels. The definition of safe levels is dependent on public health or product requirements. The choice of a sanitizer should consider the characteristics, efficacy, and applicable regulations.

The ideal sanitizer should kill microorganisms rapidly (30 to 60 seconds), providing adequate microbial reduction (about 99.9%, a 3 log reduction), and should be effective against a broad spectrum of microorganisms. It should be safe and nontoxic to employees handling it and safe for consumers at use levels. It should be acceptable to regulatory agencies. It should have no adverse effect on the product, be economical to use, be rinsable, and leave no objectionable odor or residue. It should be stable in its concentrated form and at use levels. It should be noncorrosive, compatible with other chemicals and equipment, and be readily soluble in water. Tests to detect it in solution should be easy to conduct. It should be biodegradable at concentrations expected in the waste receiving stream.

Chemical Sanitizers

Several chemical sanitizers are appropriate for various sanitization procedures within the plant. There are a variety of reference books and product bulletins that provide more information about the various chemical sanitizers and specific product formulations. Usually, experience with the sanitizers will give the only reliable benchmark for their usefulness.

Chlorine is usually provided as sodium or calcium hypochlorite. Dichloro- or trichloro- isocyanuric acid or its salts are also chlorine-releasing sanitizers. Chlorine mixed with ammonia forms chloramine, which provides a good residual chlorine source for disinfecting water. Chlorine is useful for disinfecting water, process systems, and environmental surfaces.

Iodophors are iodine-releasing formulations in a carrier. The carrier may be with or without a nonionic detergent, depending on the intended use. Iodine is useful for surface and skin disinfection. In some cases, it is useful for water disinfection but this is secondary to chlorine use.

Quaternary ammonium compounds consist of cationic surfactants. Their most frequent use is in combination with a nonionic detergent as a disinfectant-detergent or detergent-sanitizer. However, nonionic detergents usually do not clean as well as more caustic cationic detergents. Typically, one can achieve more effective cleaning and sanitizing if one separates the steps rather than combines them. This is because most cleaners operate best at high pH but most disinfectants operate best at acidic to neutral pH.

Ethyl alcohol has limited use because of flammability risks. However, it can be diluted to 50 to 70% final concentration with water partially to mitigate this concern while providing good antimicrobial activity. It is effective at cutting the film of surfactant-containing products while providing some sanitizing impact. It does not kill spores. It is useful as a hand antiseptic for short periods. Many antiseptic products are now available that provide alcohol in an emollient form to reduce drying and chapping of the hands. Ethyl alcohol is more pleasant to use than isopropyl which encourages plant personnel to want to use it in cleaning off spills and tank tops.

Phenolics may be useful when used alone. More frequently, one finds them combined with an anionic detergent for use as a disinfectant-detergent. **Formalin** is also useful. However, one should use formaldehyde in a closed system that does not permit it to escape into the air. Tighter restrictions are likely to continue because of OSHA regulations. Formaldehyde is a suspected carcinogen.

Phosphoric acid has only limited use in the plant environment. It is combined with other chemical products such as iodophors, and in tile and bathroom cleaners. **Hydrogen peroxide** is not used extensively. It is primarily useful for cleaning deep puncture wounds in skin. The bubbling action helps to lift out dirt deep within the wound, but its disinfection capacity is minimal. **Pine oil** also has limited use in a manufacturing plant.

Peracetic acid is a very strong oxidizer that is a combination of peracetic acid, acetic acid, hydrogen peroxide, and water. This is a broad spectrum disinfectant and has demonstrated good activity against biofilms. This compound does have a pungent odor and is very irritating to the eyes and nose. It is noncorrosive to stainless steel.

Physical Sanitizers

Heat is the most efficient and thorough sanitizer. In the plant, we provide it as steam (100°C/212°F) or hot water (80 to 100°C/176 to 212°F). If the steam is under pressure, the temperature can be even higher (121 to 132°F/250 to 270°F). In cosmetic manufacturing, we usually do not use dry heat. Table 3.1 defines the temperatures for dry heat and steam heat to sanitize. Note that these exposure times and temperatures are *not* those for sterilization.

TABLE 3-1

COMPARISON OF WET AND DRY HEAT TIME/TEMPERATURE (°F/°C) FOR SANITIZATION

Exposure Time	Wet Heat	Dry Heat
1/2 Hour	180°F/82°C	355°F/179°C
2 Hours	160°F/71°C	320°F/160°C
4 Hours	140°F/60°C	285°F/141°C
24 Hours	120°F/49°C	250°F/121°C

Heat has several advantages over other sanitizing agents, chemical or physical. These advantages include: penetration into small cracks and crevices, noncorrosive, nonselective to microbial groups, leaves no residue, easily measured, and inexpensive.

There are some problems with the use of steam or hot water. Heat may cause condensation problems because of the high humidity created. Only thermostable materials can withstand the heat generated by steam or hot water. High energy costs are usually offset by reduced labor costs when compared to chemical sanitizers. When they are not, we must balance the proficiency of heat treatment compared with chemical sanitizers.

Ultraviolet light provides very limited use in cosmetic microbiology. It can damage the eyes of personnel in the area if used as an area sanitizer. It has poor penetrating power and so must be close to the material under treatment. The UV intensity decreases by the square of the distance from the source. The typical use is in water sanitization. Usually, this is in combination with ozonated water systems. Plants install the UV lights as point source polishers to remove the ozone. They also kill microorganisms that may have survived the ozonation process. UV is ineffective as an air sanitizer in large, open areas.

Use of **combined chemical/physical sanitization** is the application of a cleaner or a sanitizer with steam or hot water. This approach will usually increase the effectiveness of the chemical agent. However, the chemical agent must be compatible with the heat application. For example, heat does not significantly enhance hypochlorite activity except to improve the wetting characteristics of the water. When using chlorine gas as the chlorinating agent, heat may actually drive out the chlorine. This occurs when the chlorine is not converted to hypochlorous acid, especially at low pH. This reduces the effectiveness of the chlorine sanitizer and presents a major safety hazard.

Proper Use of Chemical Agents

Chemical cleaners and sanitizers should be used only according to label directions. All appropriate employees should be thoroughly trained in the proper use of chemical agents. Improper use of an agent will reduce the effectiveness of the product and will produce unsatisfactory or less than optimal results. Overuse of a product will increase toxicity, increase corrosiveness, and may adversely affect equipment or product. In addition, it is against federal law to use an EPA-registered product in any manner inconsistent with its label. Therefore, misuse of an EPA-registered sanitizer is illegal.

Cleaning and Sanitizing Equipment

The types of equipment used to clean or sanitize production equipment and environmental surfaces include pressure sprayers, compressed air systems, and steam and foam generators. Also included are wet/dry vacuum systems, ultrasonic cleaning systems, and clean-in-place (CIP) systems. These are described further below.

Portable or Fixed Pressure Sprayers

Low-pressure systems are suitable for removing gross or loosely adhered product residues and for rinsing away cleaning solutions. High pressure systems deliver a high- velocity stream of water or chemical solution and are efficient at removing adhered residues. Low- or high-pressure systems often use chemical additives to help in removing soil.

High pressure systems operate at nozzle pressures that may be well above 1000 psi. The advantages are low water and chemical solution usage and high-pressure impingement. Some use temperatures above 82.2°C (180°F), thus adding the sanitizing power of heat. Low pressure and high volume systems may attain nozzle pressures of 400 to 600 psi, but they use large quantities of water and chemical solutions.

An ideal hose-spray system should have an on-off valve near the wand and an automatic shutoff for when the wand is not used or when the water pressure drops. This latter system is needed in steam-water mixing systems. Such systems should also have easily interchangeable nozzles that deliver droplets rather than a fog spray. The hoses should be long enough to do the job and strong enough to withstand use-pressures, temperatures, chemicals, and external abrasion. The system should have premixing mechanisms to provide inexpensive use of chemicals and water and be composed of internal parts resistant to corrosion and scaling. It should have enough capacity to maintain use-temperatures and pressures, provide low maintenance costs, and be reliable over several hours of continual use. Finally, such systems should have a quiet operation and adequate safety features for employees.

No system will combine all the features, but one should rate any system against them. A centralized or decentralized fixed system with lines running to fixed, multiple locations throughout the plant is preferable to multiple portable units.

Portable or Fixed Steam Generators and Sprayers

With these steam generators, one can deliver high-velocity steam or a spray of steam with or without chemical additives. They provide a very effective removal of product residues, films, oils, and fats. They also have some sanitizing effect. Steam can be very hazardous to workers. Personnel must use it with extreme caution. The units may be centralized or decentralized fixed systems or portable units.

Portable or Fixed Foam Generators and Applicators

Foam generators provide a stream of chemical foam at ambient temperature under air pressure of about 100 to 400 psi. Foam provides a longer contact time between the cleaning agent and the soil. The generators reduce the required amount of cleaning solution. It prevents splashing of cleaning solution. Thus, it also reduces irritating fogs and vapors. Foams help clean the undersides and other hard-to-reach surfaces of equipment by clinging to surfaces. Finally, they show the employee the equipment areas covered.

A foam's suitability for a particular job depends on the formulation, wetness, and contact time. Foam systems may be permanent or portable and need a compressed air source. Foams are not suitable for removing attached soil that requires turbulence for shear cleaning.

Compressed Air Systems
These deliver compressed air at work stations. One uses this air to blow away product residues, remove excess cleaning solutions, dry surfaces or components, and aid in applying foam. Compressors should have air filtration devices. Air compressors may be centralized, decentralized, portable, or fixed. The risk of creating aerosols and thus spreading the material throughout the plant should be balanced against its usefulness.

Wet-Dry Vacuums, Portable or Fixed
One uses wet vacuum systems to remove excess or pooled cleaning solutions and water from equipment, floors, and other surfaces. Dry vacuum systems are necessary for processing dry products or ingredients. They are necessary where wet methods cannot be used, and for dry removal of residues, powders, and soil. Most plants centralize their vacuum systems. However, they may be portable as well.

Ultrasonic Cleaning Systems
These consist of an immersion tank, a cleaning solution at 65.6°C (150°F), an ultrasonic generator, and ceramic transducers. The ultrasonic generator produces waves in the frequencies of 30,000 to 40,000 hertz (cycles per second). The ceramic transducers are for converting the ultrasonic energy into mechanical vibrations. Ultrasound is useful for cleaning small pieces of equipment that are delicate or hard to clean.

Recirculation Wash Tanks
These tanks consist of an immersion tank in which cleaning solution recirculates across and through the tank by jets at elevated temperature. The combination of cleaning agent, hot water, and the physical shear force of circulation removes soil. This removal is mainly from equipment pieces small enough for immersion into the tank.

Automatic Washers
These consist of cabinet or tunnel washers that can pre-rinse, wash, rinse, sanitize, or dry items, and can warm or cool items, if desired. They are often used in recycling containers for product fills.

Clean-in-place (CIP) Systems
Clean-in-place systems depend on the high velocity circulation of proper cleaning and sanitizing solutions. These solutions are circulated for a specified time at a specified temperature in a closed system designed for that purpose. No dismantling of equipment is necessary. A CIP system

can apply to an entire production line or to individual parts of the line. Clean-in-place systems are useful for product-carrying lines, fittings and valves, tanks, production vessels, centrifugal pumps, heat exchangers, evaporators, and conveyor belts.

The advantages of CIP include reduced labor costs, automation, and the ability to use strong alkaline cleaners (1000 to 1500 ppm active alkalinity). Other advantages include faster cleanup, fewer leaks, and less damage from frequent disassembly that is otherwise necessary. Coupled with the ability to recirculate the cleaning solutions (thus adding a cost savings), these advantages add up to a better and more cost effective cleaning job.

Miscellaneous Equipment and Supplies

Manual or special cleaning of equipment or its parts may require special types of equipment or supplies. These include scrapers, nonmetallic abrasive pads, squeegees, soft or stiff bristle brushes, brooms, wiping cloths, buckets, sinks, pallets, ladders, and stools. Wire brushes may scratch and damage equipment. Steel wool or metal scouring pads can shed metal particles that can become incorporated in the product as well as scratching equipment. Never use wire brushes, steel wool, or metal scouring pads on production equipment, especially on the surfaces that contact product. Even minute scratches barely visible to the eye represent Grand Canyons in the microbial world ... perfect places for growth.

Cleaning and Sanitization Procedures

Cleaning and sanitization frequencies, procedures, products, and equipment will depend on the area or the equipment being cleaned and sanitized. Each plant should establish written cleaning and sanitization procedures and frequencies. These procedures must be written. Continuous monitoring by supervisory personnel is necessary to check the adequacy of schedules and procedures. This also allows data-based decisions for timely remedial action, revision of procedures, or adjustment of schedules.

The three areas to consider when establishing these procedures are the warehouse, manufacturing and production areas, and manufacturing and filling equipment. In addition, one must establish appropriate monitoring, training, and waste disposal practices.

Warehouse Areas

Warehouses are used to store raw materials, packaging materials, finished products, equipment, and miscellaneous supplies. Environmental conditions within the warehouse must be sanitary to prevent microbial contamination of raw materials, finished products, and the environment.

The following actions are necessary to maintain acceptable environmental conditions in warehouses.

Insure that janitorial personnel receive proper training to maintain acceptable environmental conditions.

Keep aisles neat and clean by sweeping or vacuuming followed by damp mopping or machine scrubbing as needed.

Clean up spills promptly using methods appropriate for the type of spill.

Store all materials under clean conditions, in an orderly fashion.

Protect the raw materials from contamination.

Properly identify all raw materials.

Assure that all containers are clean and kept in a clean, segregated area before they enter the manufacturing areas.

Clean the exteriors of containers before transporting materials into manufacturing areas.

Maintain routine insect and rodent (pest) control programs to avoid vector-borne microbial contamination.

Inspect warehouses routinely to determine the adequacy of sanitation and pest control programs.

Plant Manufacturing and Production Areas

Each area of the plant requires specific cleaning and sanitization procedures and schedules. The procedures depend on the type of activities conducted there. Each area requires specific needs. There are, however, some general guidelines. Insure that personnel receive proper training in cleaning procedures and use of chemical products.

One must clean the walls, ceilings, pipes, and fixtures on an appropriate schedule. This may be monthly, quarterly, or whenever visibly soiled, but must be included in an SOP that is written. Cleaning can be by vacuuming to remove dust and loose debris. However, one must not generate airborne dust during cleanup operations. The surfaces may need to be wet cleaned occasionally, but use caution in spraying water and cleaning agents everywhere since the goal is to keep the manufacturing area as dry as possible to limit microbial growth. Use appropriate wall or ceiling cleaning equipment, or pressure sprayers only when necessary.

Clean floors daily by sweeping or vacuuming followed by damp mopping or machine scrubbing. One also may apply the right sanitizer when needed. Clean up spills promptly using the right cleaning method for the type of spill. Maintain all cleaning equipment in a clean and sanitary condition. Wherever possible, provide separate equipment for each area.

Manufacturing and Filling Equipment

Always prepare and document a schedule for cleaning and sanitizing each separate piece of equipment. As with the plant itself, cleaning and sanitizing procedures must be specific for each type of equipment. However, there are some common basic cleaning and sanitization practices.

Thoroughly clean and sanitize all compounding and filling equipment between batches of products and after operations. Determine cleaning frequencies for continuous process or consecutive batch equipment by appropriate testing.

Clean manufacturing and filling equipment by soaking, by spraying, by CIP, or by any methods that use appropriate chemical agents and cleaning equipment. Always use the least labor-intensive cleaning procedures suitable for the task provided they are effective. Thoroughly rinse all cleaning solutions from equipment surfaces using an acceptable water supply. Properly protect cleaned and sanitized equipment to prevent contamination.

Monitoring

Supervising personnel should monitor work practices and compliance with established procedures and schedules. Quality assurance personnel should check the adequacy of the cleaning and sanitizing processes by visual observation and by appropriate microbial sampling. Document all the monitoring that is included in GMP compliance.

Training

Training of all personnel who perform cleaning and sanitizing tasks must be thorough. This training should include the proper use of the chemicals, equipment, and procedures for each task. They should be thoroughly familiar with written procedures used for each piece of equipment for which they are responsible. This training should include how to disassemble equipment as needed.

All employees who handle chemical agents, or may be in areas of their use, must receive hazard communication training according to the OSHA Hazard Communication Standard (worker-right-to-know). Address all the requirements of the standard and document all training.

Waste Disposal

Place all refuse in covered leakproof, rustproof containers lined with a plastic insert liner. Empty all refuse at least daily, more frequently if needed. Upon emptying, clean containers as needed before reuse. Store refuse waiting for disposal in a protected area maintained in a sanitary condition consistent with a pest control program.

Dispose of product or process wastes and all other wastes generated in the plant according to applicable local, state, or federal regulations. Train waste handlers in the proper handling and disposal of waste.

Handling should be according to plant policies and procedures, and regulatory requirements.

PERSONAL HYGIENE

Good personal hygiene by employees is a significant means of controlling microbial contamination in the plant. Personnel who do not practice good personal hygiene will nullify all the work just described above. This happens despite having the best facilities, equipment, and processes, and good cleaning/sanitization practices. Personal hygiene is also the one microbial control factor that is the least controllable by the supervisory and quality assurance personnel. Therefore, one must provide the workers with proper training in good personal hygiene habits. We must also motivate them to want to practice good personal hygiene habits. Behavioral changes must occur where necessary. The most effective process to encourage cleanliness is education. There are several basic hygiene practices to observe.

Handwashing

Personnel must always wash their hands after using the bathroom or whenever they are returning to the production area. The hands should be clean before working on any product or production equipment. Prominently post signs to remind employees to wash their hands after using toilet facilities and before returning to work.

Proper handwashing requires readily accessible and adequate hand washing facilities that encourage handwashing. Facilities should include hot and cold running water, a pleasant to use hand cleansing agent, and single-use towels or noncontact hand dryers.

Use of an antimicrobial hand cleansing agent is not as important as using one that encourages the hand washing process. The agent should be pleasant. It should feel good during use. It should impart a soothing smooth texture to the hands after use and leave a pleasant odor to the hands that lingers. These signals will promote a behavior modification process that results in increased hand washing. Use of an emollient hand cleansing agent will also help prevent skin irritation, chapping, and cracking that can discourage handwashing. This cracking also provides tiny foci for microbial growth. Hot air dryers also may cause excessive skin drying and cracking.

If the plant uses hand-dip basins or antiseptic sprays, locate them immediately inside the reentry site to the production area. Select the antiseptic agent used for these rinses based on both antimicrobial activity and mildness on the skin.

Dining/Smoking

No food should be eaten or allowed in production areas. Provide dining facilities and break areas close to work areas. Maintain them in a good, sanitary condition. Provide ample trash disposal to encourage proper use of the facilities. Handwashing facilities should be near these areas. Cleaning items for spills should be available. Clean and disinfect the dining area at least daily.

If smoking is allowed in the plant, permit it only in designated smoking areas. Do not permit tobacco products in production areas.

Wearing Apparel

Wear only authorized apparel in the plant. Whether wearing personal clothing or company-provided uniforms, they should be clean when the shift starts. Generally, one should cover street clothes with lab coats or other clean cover-ups. Follow all plant policies for the wearing of coats, smocks, aprons, and jumpsuits. These work clothes should never be worn out of the production area and should be changed as often as necessary.

Wear hair covering (hats, beard covers), masks, or gloves correctly, and change them as often as needed. Production personnel should not wear loose jewelry, earrings, brooches, high-crowned rings, or wrist watches. Wearing of plain-band rings (such as wedding bands) is permissible; fingernail polish or long nails are not permissible.

The personnel handling the product or working with production equipment should not wear badges, decorative buttons, or identification cards. If they must wear identification badges, they should be securely attached so they do not accidentally fall into the product or equipment.

Personal Habits

Around product or production equipment, people must not scratch or place their fingers in or around their nose or mouth. They should not sneeze or cough on product or equipment. Watch for any other personal habits that could contaminate the product or the equipment.

RAW MATERIAL HANDLING

Raw materials do not have to be sterile for use in cosmetics. Proper handling is important to reduce the potential for microbial growth in the raw material and to prevent introducing microorganisms into the raw material.

Upon receipt of raw material, be sure to inspect package integrity. Bags should not be torn or wet. Drums should not be rusty or have too many dents. If you accept materials that are not in good condition, segregate them into an area for special handling.

Place raw materials that are sensitive to microbial contamination into a segregated quarantine area until released by the microbiology laboratory. Once the raw material is proven acceptable, remove it from the hold area and place it into the warehouse for use by manufacturing. Store all raw materials in an orderly manner so date codes are visible for proper rotation.

Sample incoming raw materials in an aseptic manner. This reduces the potential for introduction of microorganisms into the material that may result in contamination. Train the manufacturing personnel in the proper method of opening containers and handling raw material. Clean all the containers before opening. Properly reseal partially used containers of raw materials to prevent contamination during storage. Repackage damaged containers of raw materials before returning them to storage.

SANITARY DESIGN OF EQUIPMENT

Sanitary equipment design is the most neglected concern when developing or purchasing equipment. Equipment can be one of the major sources for contamination of products during manufacturing. Properly designed machinery in operation will release microorganisms from locations harboring biofilms. These areas are usually not accessible for proper cleaning and sanitizing. Processes like milling, sieving, centrifuging, and mixing may release contaminants into the environment. These contaminated aerosols can then transfer to other areas of the plant to cause contamination of other products.

The microbiologist plays an important role in determining the proper cleaning and sanitizing procedures. He or she should also play a central role in approving sanitary design of the equipment before its purchase or construction. He or she should consider the materials used to make the equipment, installing sanitary fittings and valves and providing appropriate slopes for proper drainage of the equipment components.

Detergents and sanitizers, which may be alkaline or acid, may affect the internal equipment surface. Heat sanitization may have detrimental effects on the construction materials too. Some of the effects are swelling of gaskets, expanding pistons and then freezing them into place, and shorting out electrical circuitry.

The microbiologist should consider equipment design that permits ease of cleaning and inspection. This will help reduce critical hazard areas. Some of these key areas are dead-legs, poor drainage, and pipes that slope back into a pump or tank.

The microbiologist should see that construction of process equipment and systems is done with materials that are compatible with the product. The system should also withstand the temperatures and pressures incurred during the cleaning, sanitation, and manufacturing process. The most common material used to construct process equipment is stainless steel of at least type #304; #316 is ideal. Surfaces should be ground smooth along welds and joints. Materials to be cautious of are cast materials because of surface roughness, aluminum because of reactivity, and plastic because of porosity.

The cleanability of the equipment is very important. The people working with the equipment must be able to disassemble it easily. This ability allows for proper cleaning. It prevents cross-contamination of products from previously run products, and it helps in controlling microbial growth. Surfaces need to be easily accessible for inspection and mechanical cleaning. Equipment should be located in an area that provides for easy cleaning and maintenance and kept in areas that are sanitary to avoid possible cross-contamination.

Compounding equipment must be of a sanitary design from the water inlet through the outlet valve of the mixing tank. It should have covers to prevent airborne contamination and rounded bottoms to promote proper drainage. Process control equipment such as flow meters, thermometers, pressure gauges, viscosity, and level meters need special attention to prevent product buildup.

Pumps used to move liquids to various locations or to homogenize or recirculate the product should be easily accessible for dismantling and cleaning. The inspection of pumps should include bearing and shaft seals for leaking or product retention, and gaskets and impellers for proper integrity. The pump should empty completely. Screw-threaded connections should be avoided as these may be a source of contamination.

Transfer hoses should be of a material that is compatible with product and cleaning agents. Hoses should have sanitary fittings and be as short as possible. This helps in hose maintenance. Clean hoses and hang them properly to drain when not in use. After the hoses have properly drained and dried, cover or cap the ends to prevent contamination after cleaning.

Transport pipelines should be made with stainless steel and fitted with sanitary (or dairy) fittings. Other materials used in transfer piping have shown the potential for biofilm buildup and are difficult to clean. All the pipelines should be sloped to drain properly. Horizontal sagging lines prohibit proper cleaning and may become sites of product buildup. Avoid dead-ends, "air-hammer" pipes, right angles, and vertical bends that become contamination reservoirs. In-line filters may become a source for contamination if not cleaned routinely. Filters should be designed for easy cleaning, sanitizing, and inspection.

Filling equipment used today in the cosmetic manufacturing environment is very diverse. There are certain common important features that are critical in filling equipment design. These features allow for proper cleaning and sanitizing. The equipment should be designed for fast and easy dismantling. Detection switches that are sanitary should be used to determine proper product levels in filling equipment hoppers. Eliminate dead-ends in product transfer piping. Include bacteriological filters and air line dryers to prevent air line condensates from getting into the product. Monitor all critical control points to reduce contamination of product at the filling process.

Product containers are another item associated with filling equipment that needs monitoring. Protect them from environmental dust and moisture. Cover any portable tanks and product hoppers that contain product while in use.

SANITARY DESIGN OF THE PLANT

The design and construction of the plant will directly influence the ability to maintain clean and sanitary conditions within the plant and outside the plant. The design and construction of the plant should take into account the environmental surfaces, drainage, location of pipes and utility systems, and pest control. It should also be convenient for traffic and material flow, proper lighting, and cleaning. Finally, the entire plant should be kept in good repair to meet the sanitary design parameters.

Environmental Surfaces

The environmental surfaces (floors, walls, ceilings, and windows) should limit the deposition of dust and debris and facilitate cleaning and sanitization.

Floors

Floors should be smooth and nonporous. They should be easily cleanable and withstand the traffic load that they will bear without sustaining damage. Vinyl flooring is easily cleanable but may not withstand heavy traffic. Concrete, quarry tile, or terrazzo flooring may be more porous but they are sealable with a durable sealant, such as epoxy coating. This makes the floor readily cleanable and protects against the corrosive action of cleaning agents. The subflooring over which any flooring lies should be stable to prevent cracking of the flooring material. Since cracking interferes with floor cleanability, any cracks that appear should be sealed or the flooring replaced.

Floors should have a slight slope leading to drains. These drains should be constructed and maintained so they will remain unclogged and will not backflow. They should dry out completely between exposure to water without creating traps of dried materials. They should not create odors. Keep floor drains clean to prevent cross-contamination. All sink and plumbing fixtures should remain clean and in good repair. Install back siphonage prevention devices. Create a rounded concave cove along all junctions between floors and walls to restrict the buildup of debris and to ease cleaning. The flooring material should extend up the wall for a short distance.

Walls, Ceilings, and Windows

Construct all walls and ceilings with smooth, nonporous, easily cleanable materials. These materials should be resistant to the corrosive action of cleaning agents. Often used wall coverings are vinyl, durable paints, and ceramic tile. All seams and joints found in walls and ceilings should be sealed. Any cracked, damaged, or peeling wall and ceiling surfaces should be repaired or replaced as soon as possible. Use kick plates or wall shielding to prevent damage where doors, corners, or walls are subject to contact from heavy equipment.

Projections on walls and ceilings should be kept to a minimum to maintain a smooth, easily cleaned surface. Lighting fixtures and air vents to the ceiling should be flush with the ceiling surface. Keep suspended piping from the ceiling to a minimum. Locate piping behind a ceiling overhead where possible. However, any piping or utility systems should be readily accessible and cleanable.

Windows should be nonopening and flush with the wall surface. Avoid window ledges because they are subject to dust and moisture deposition. When using hot water or steam in cleaning/sanitizing, or where the humidity is high, installation of thermal windows reduces moisture condensation.

Traffic and Material Flow

Design of the plant should provide a one-way, sequential flow of materials. Raw materials should enter the manufacturing area from one set entry area. Any materials taken into the manufacturing area should be clean. The outer secondary packaging materials should be removed before they enter either the raw materials dispensing area or the manufacturing area. This one-way sequential flow of materials will restrict any cross-flow of clean and contaminated materials. It adds some assurance the right sanitation and product quality standards are met.

Similarly, we must also apply restrictions to personnel traffic. Do not allow unnecessary traffic in the plant. Provide separate pathways. Only authorized personnel should be allowed in the manufacturing area. Enforce restricted access through set entryways. Designate a location to put on any special protective apparel worn by plant personnel.

Lighting

Provide adequate lighting in all plant areas particularly around machinery. Sufficient lighting allows personnel to accomplish their assigned tasks in a safe manner and to meet performance requirements. Applicable OSHA lighting requirements and illumination standards should be met. Install safety lights where flammable products are present and in areas of water contact. Install them wherever lights are subject to contact with water.

Pest Control

Integrated pest control is a combination of sanitation first and the use of pesticides second. Sanitation in this case includes the design of the plant to control pests. Design and construct all areas of the plant to restrict access to insects, rodents, birds, or other pests. We have already discussed the design and construction of the plant to provide for minimum debris buildup and maximum cleanability. Here, we will discuss how to control insect, bird, and rodent pests.

All unnecessary openings into the plant should be sealed and all required openings to the outside screened. Keep doors closed when not in use. Use loading dock doors or closures suitable for loading dock operations. Vertical air curtains can be used to restrict the entry of flying insects through doorways where doors are open for a lengthy time. All harborage areas inside the plant must be eliminated. Pay special attention to warehouse and storage areas not subject to the strict cleaning and sanitization practices required in the production areas. Also, correct any conditions on the exterior of the plant that serve as attractants or harborage for pests.

Plant Exterior

Keep plant exteriors clean and orderly. Control excessive growth of weeds and decorative plants to restrict insect or vermin harborage. Maintain waste storage areas in a sanitary manner to control pests and for aesthetic reasons. Clean all spills promptly. The building exterior must be kept in good repair to help control environmental conditions within the plant and restrict pest entry. The roof must have good drainage and be kept in good repair to prevent leakage of water into the building.

Good drainage must be provided around the plant to eliminate standing water that could serve as a breeding place for insects or could allow prolific microbial growth. Drainage must also accommodate runoff from rain or melting snow to avoid flooding and to allow for spill cleanup.

Dike all bulk chemical storage areas that are above the ground to contain spills. Construct below-ground chemical storage areas so they will be impervious to contaminating ground water. Provide drains if the spilled chemicals can be safely discharged to the sanitary sewer or the storm drain.

SANITARY DESIGN OF WAREHOUSES

The Federal Drug and Cosmetic Act states the general requirements for the design and construction features of storage areas. These regulations set the minimum requirements for Current Good Manufacturing Practices. Products not stored according to current good manufacturing practices (cGMP) may become adulterated. Therefore, warehouses must provide conditions that will not have harmful effects on the contents stored within them.

Sanitary design includes the building and the outside grounds around the warehouse. Grounds surrounding the warehouse should have good drainage to prevent the breeding of insects. Removal of discarded equipment, lumber, litter, waste, and weeds is necessary to prevent breeding and harboring of insects, rodents, and other pests. Place outside waste disposal containers in a well drained area. Keep them clean and in an operable condition. Keep them covered between uses.

Design the warehouse walls, ceilings, and floors with finishes resistant to normal use and accessible for maintenance and cleaning. The cleanability of a surface is an important microbiological aspect. Soil and dirt accumulation on a surface has the potential of becoming airborne. A clean, dry surface will not retain dirt particles under normal use conditions compared to a surface that has not been cleaned. These uncleaned surfaces will lead to the release of contaminated particles into the warehouse environment.

Since adequate lighting is needed to help the handling, processing, and examination of the storage materials, good lighting must be provided in the warehouse to permit inspection, cleanup, and repair of the building and structure. All the equipment used in the warehouse, such as trucks, jacks, and dollies, must be cleanable. The equipment must not contaminate materials or product with lubricants, metal fragments, or water. The ventilation must be adequate to prevent condensation from developing on ceilings, fixtures, ducts, pipes and the product itself. Other sanitary design features include sanitary plumbing to provide enough potable water throughout the warehouse and to properly remove sewage and wastes from the warehouse. Locate washing and toilet facilities in areas that will not contaminate warehouse contents.

SANITARY DESIGN OF WATER SYSTEMS

It is critical to design water systems that permit effective sanitization. Holding tanks and associated piping should be stainless steel. Avoid using pipes made of copper or galvanized piping. Never use PVC or black iron pipe. Water lines entering mixing tanks should not permit back siphonage of the tank contents into the water system. This is easily done by providing a shielded air gap. Prevent back siphonage in the system when pressure drops in the water line by avoiding cross connections. Construct filters in the water system in parallel. This allows a filter needing service to be isolated from the rest of the system.

Supply water treatment systems with chlorinated city water. This treated water should go directly into the stainless steel storage tanks and be periodically heated to 180°F. Circulate the heated water through the system and into the storage tank. An alternative to heat treatment of water is ozonation. Use ultraviolet lights at the point of use to remove the ozone and further sanitize the water.

SANITARY DESIGN OF AIR SYSTEMS

The design of air systems must be specific for each area served. It should also consider the air quality needs for the operations performed in that area. This will require several different air handling systems based upon the air quality needs in the areas served. These systems are commonly known as HVAC (heating-ventilation-air conditioning) systems. The HVAC system design must consider several aspects. These aspects include the quality of the incoming air, temperature, humidity, air exchange rate, and desired air purity. Also, consider the location of incoming/exhaust air vents or ducts, and the duct work for the control of airflow patterns.

Air Filtration

Filter all incoming air supplies. The degree of filtration required depends on the air quality desired in the area served and the quality of the incoming air. Raw material areas may require low efficiency air filtration while manufacturing/filling areas will usually require more highly purified air and air filtration. Most systems require sequential filtration with a primary filter providing lower level filtration, and secondary and tertiary filters providing higher level filtration. Install filters downstream from the air handling units (fans). Electrostatic precipitators are sometimes useful in place of a final filter to remove fine particulate matter. Servicing filters or changing them on a regularly scheduled basis is preferable to changing them only when they are dirty and restrict air flow (as determined with a manometer).

Air Treatment

Many plants need to either humidify or dehumidify the air. This need depends on area requirements, the time of the year, or the geographical location of the plant. Maintain such systems in a clean and sanitary manner. The water in the humidification system and the dehumidification collect pans and drains must be monitored and properly maintained to prevent the proliferation of microorganisms.

Air Exchange

Balance the air supply systems to provide proper air exchange rates for the areas served. It is preferable to have cleaner areas under slightly higher air pressure than adjoining dirtier areas. The positive pressure in the clean area will restrict the flow of contaminated air from dirtier areas via openings where air exchange occurs. Unserviced filters become overloaded and restrict incoming airflow, thereby lowering the planned air exchange. Any nonessential or unplanned openings, such as cracks in walls can alter the exchange rate. This makes good building maintenance essential to maintain the desired airflow.

Energy Conservation

Some plants turn off the HVAC systems at night, on weekends, or holiday periods for energy conservation purposes. When this is done, one must turn them on again in enough time to bring the air quality back into conformance with requirements. This will usually require a few hours, depending on the size of the area and the efficiency of the air handling system. In some critical areas, or where there is almost continuous production, it will be necessary to leave the system on full operation.

Makeup Air

Some HVAC systems use 100% makeup air. Many HVAC systems use recirculated air or provide only a partial fresh air makeup supply. Locate air intakes for the incoming air supply away from the exhaust air ducts or loading docks. Keep them away from any other area where chemical air pollutants may enter the fresh air makeup supply. The usual filtration used in HVAC systems does not remove chemicals from the air.

Monitoring

Microbial or particulate air monitoring and simple airflow measurements are usually within the capabilities of plant personnel. However, the services of a ventilation engineer or industrial hygienist will permit more intensive and extensive evaluation of HVAC systems.

SUMMARY

Keeping the plant sanitary is one of the critical preventive measures to mitigating contamination of product. In fact, in a well-controlled, sanitary plant where all raw materials are carefully controlled and where equipment is kept industrially sterile, even marginally preserved products can be manufactured.

The book by Troller[1] listed below remains a classic reference for sanitation and design of a consumer products manufacturing plant.

REFERENCE

1. Troller, J. A. 1983. *Sanitation in Food Processing*. Academic Press, Orlando, FL.

SECTION TWO: COSMETIC MICROBIOLOGY TESTING METHODS

4 PRESERVATIVE EFFICACY, MICROBIAL CONTENT, AND DISINFECTANT TESTING

Scott V.W. Sutton, Mary Anne Magee, and Daniel K. Brannan

CONTENTS

0-8493-3713-5/97/$0.00+$.50

INTRODUCTION

Cosmetics do not need to be sterile, but they must be adequately preserved. When consumers use cosmetic products, they repeatedly challenge the cosmetic with microorganisms in saliva, on dirty hands, and in tap water. Microbial growth may occur in cosmetics and toiletry articles kept in the bathroom and subjected to heat and humidity.[1] These products include mascara, eye shadows, shampoos, facial powder, and facial lotions (foundations and moisturizers).

Cosmetics intended for use in the eye area are of particular concern. The cornea, especially if compromised, can be extremely vulnerable to infection. There have been several instances of mascara contamination from *Pseudomonas aeruginosa*.[2] Microbes in products can result in infection, discoloration, production of gas, or odor formation. Typical contaminants of cosmetics include *Enterobacter* spp., *Klebsiella* spp., *Serratia* spp., and *Pseudomonas* spp., the major contaminants.[1,6,7]

Although cosmetics occasionally are contaminated with spoilage microorganisms, the biggest threat of contamination is the presence of pathogens that pose a health threat.[4] However, even nonpathogenic spoilage microbes in a cosmetic may cause disease under appropriate conditions. For example, a contaminated cosmetic may be invasive if one applies the cosmetic to cover a blemish or break in the skin.[5] With the rise of immunocompromised individuals in the population due to the pandemic of AIDS, the problem becomes even more acute.

The microbiologist uses a variety of chemical preservatives to prevent contamination by pathogens or spoilage microorganisms. This use extends the shelf life of the product. The preservatives include benzyl alcohol, boric acid, sorbic acid, chlorhexidine, formaldehyde, parabens, quaternary ammonium compounds, phenol, imidazolidinyl compounds, and others. One may reference the excellent reviews on this subject.[1,8–10]

PRESERVATIVE EFFICACY METHODS

Test Methods Currently Followed

There are three sources for testing guidelines of preservation efficacy in a cosmetic or toiletry product. These sources include the Cosmetic, Toiletry, and Fragrance Association (CFTA), the American Society for Testing and Materials (ASTM), and the U.S. Pharmacopeia (USP). All three involve challenging the cosmetic formulation with microorganisms. However, specific differences among the procedures occur to address the concerns of the parent organization.

The Food and Drug Administration has revised the Microbiological Methods for Cosmetics, Chapter 23 in the Bacteriological Analytical Manual.[5] A preservative challenge test that preferably is predictive of consumer contamination is FDA's desire. They have not yet released the final version of such a method.

General Procedure of Preservation Efficacy Testing

Each of the methods has specific requirements about recommended organisms, media, growth and storage conditions (see Tables 4-1, 4-2 and 4-3). Standardization of all the requirements of the challenge must be done to provide reproducible data.

The preparation of the challenge organisms is of special importance. The growth and preparation of the challenge organism determines the physiological state of the cell. This state has a direct influence on the results of any assay of disinfection efficacy.[11,12] One must maintain cultures of microorganisms which are subcultured on appropriate media to assure viability and resistance. Standard conditions for organism preparation and storage are essential for reproducible results. Biocidal tests do not use individual cells, but populations of cells. The data generated from these tests is less variable if the cell population is homogeneous. Liquid cultures, or confluent growths on solid media, are adequate for the reproducible growth of inocula, but there are conflicting results as to which provide the most reproducible results.[12] Anytime growth is slowed due to stress (including exposure to biocides), or because of some nutrient limitation, microbes become more resistant to the biocide.[12a]

Bacteria used as an inoculum are usually at a concentration of 1.0×10^8 colony forming units per milliliter (CFU/ml). This allows for a 1:100 ratio of inoculum to product dilution, giving a final recommended concentration of bacterial challenge of 1.0×10^6 CFU/ml or gram. The high initial concentration of the challenge organism reduces the dilution of product upon inoculation. One typically uses fungi and yeast at a final concentration of 1.0×10^4 fungal spore or yeast/ml or gram (CTFA). The USP recommends 1.0×10^5 fungal spores or yeast/ml or gram (ASTM).

TABLE 4-1

SUGGESTED ORGANISMS FOR TESTING

Organism	ATCC	CTFA	ASTM	USP
Bacteria				
Gram Negative				
Enterobacter aerogenes	13048		+	
Escherichia coli	8739			+
	na	+		
Proteus spp.	na	*		
Pseudomonas aeruginosa	na	*		
	9027		+	+
	15442 or 13388	+		
Gram Positive				
Bacillus subtilis var. *globigii*	na	*		
Staphylococcus aureus	6538	+	+	+
Staphylococcus epidermidis	na	*		
Yeasts/Fungi				
Aspergillus fumigatus	na	*		
Aspergillus niger	16404		+	+
	9642	+		
Candida parapsilopsis	na	*		
Candida albicans	10231		+	+
	na	+		
Fusarium solani	na	*		
Eupenicillium levitum[1]	10464		+	
Penicillium luteum	9644	+		

Note: na, an ATCC number was not cited in the method.

[1] Name given in table is ATCC name. The name listed in the method is *Penicillium levitum.*

* CTFA Preservation Testing of Aqueous Liquid and Semi-Liquid Eye Cosmetics.[13]

The USP calls for standardization of the yeast/fungi inoculum to 1.0×10^8 yeast/spores per ml. This translates to an initial inoculation of 1.0×10^6 microorganisms per milliliter or gram of product (see Table 4-2). Note that the dilution of the test product should not exceed 1%.

The challenge inoculum for the test consists of inoculating a known amount of test sample with the test organism. The technician then removes aliquots from the test sample and dilutes them in a neutralizing broth. Neutralization of the preservative must occur in the broth to avoid inflated estimates of efficacy. In these procedures, some residual preservative will carry over into the recovery medium. One should use a neutralizer to inactivate this residual and permit the organisms to grow. The CTFA and the ASTM both address the neutralization of the preserved test solution.

One determines the number of surviving colony forming units by plating the broth dilutions on the proper agar. Table 4-3 lists several of these media and their uses. One then incubates the inoculated plates at the optimal temperatures until the colonies grow large enough to be countable. The number of surviving CFU's are then calculated from the dilution giving countable plate growth (traditionally 30 to 300 CFU/plate).

CTFA Method[13,14]

We provide a listing of the challenge organisms suggested for cosmetics in Table 4-1. This listing also includes those organisms associated with the eye. These organisms may be drawn from indigenous microflora around the eye, clinical isolates, or isolates from contaminated products.

One should culture the organisms on suitable media. Recommended media for bacteria include Nutrient Agar,[14] Tryptic Soy Agar,[15] or Eugonagar.[16] Culture media for fungi include Sabouraud Dextrose Agar,[17] Potato Dextrose Agar,[18,19] or Mycophil Agar.[20–22] Once growth occurs, one may maintain bacteria and yeast at 5°C on slants. One effective means of maintaining fungi is to store them at room temperature on slants. Weekly or periodic subculturing can be done to assure the viability of the microorganisms, but this risks loss of resistance. Cultures may also be frozen or lyophilized as an alternative to permit the stability of the microorganism and end frequent subculturing. The main advantage to these latter storage means is that they prevent loss of genetic resistance factors. These factors, as well as ones requiring phenotypic expression, are sometimes lost with frequent subculturing in media without the selective pressure of the biocide to maintain resistance due to genotypic and phenotypic characteristics. One may also maintain contaminating microorganisms in the same marginally preserved products from which they were isolated. This practice maintains continued resistance to the preservatives.

To prepare the inoculum, the bacteria and yeast may be grown in broth or on agar. The harvested culture may be used to challenge the test sample. The microorganism challenge is usually done in at least 20 grams or ml product. A recommended challenge level is 1.0×10^6 bacteria or 1.0×10^4 fungi and yeast per gram product. The CTFA prefers use of single cultures for the challenge or pooling of similar organisms to provide specific data for each organism or categories of organisms.[23] The inoculum volume should not be more than 1% of the final volume. These challenge levels represent a larger microbial challenge of the product than might be expected from normal consumer use. Therefore, most challenge tests are far more stringent than they need to be in order to insure that the product withstands contamination during consumer use. This is especially true when we consider that many products also have protective mechanisms for delivering the product without it coming into

TABLE 4-2

COMPARISON OF THE THREE CHALLENGE SYSTEMS

	CTFA	ASTM	USP
Detailed passage information	No	Yes	Yes
Harvesting solution	na	Ster.water	Ster.saline
Use of actual contaminants	Yes	No	No
Specifics for fungal harvesting	No	Yes	No
Fungal harvesting solution	na	na	Polysorbate 80
Standardization procedures	No	Yes	Yes
Mixed culture inoculum	Not recommended	Yes	No
Pure culture inoculum	Yes	Permitted	Yes
Standardized inoculum CFU			
Bacteria	na	1×10^8/mL	1×10^8/mL
Yeast/fungi	na	1×10^7/mL	1×10^8/mL
Challenge CFU—bacteria	1×10^6/gm or mL	1×10^6/mL	10^5 to 10^6/mL
Challenge CFU—yeast/fungi	1×10^4/gm or mL	1×10^5/mL	10^5 to 10^6/mL
Inoculum amount	≤1% final vol	1 mL	0.1 mL
Sample size	20 gm or mL	100 gm	20 mL
Sample times (days)	0,1,2,7,14,21,28	0,7,14,21,28	7,14,21,28
Use of neutralizers addressed	3 general broths	Separate method E 1054-85	No
Neutralizer evaluation	No	Yes	No

Controls	No	Yes	No
Preservative Effectiveness			
Bacterial Reduction	7 days 99.9% reduction	7 days 99.9% reduction	14 days 0.1% remaining
Fungal Reduction	7 days 90.0%	28 days 90% or no increase during test period	14 days same or below initial concentration

Note: For microorganisms, see Table 4-1.
na, not applicable.

* CFU per·ml or g of product.

TABLE 4-3

VARIOUS MEDIA USED IN PRESERVATIVE EFFICACY TESTING (PET)

AGAR	Microorganisms	Purpose	Procedure	Ref.
Tryptic Soy Agar	Bacteria	General	CTFA	15
Nutrient Agar	Bacteria	General	CTFA, ASTM	14
Eugonagar	Variety	General	CTFA	16
	Yeast/fungi			
Sabouraud Dextrose Agar	Fungi	General	CTFA	17
Potato Dextrose Agar	Yeast/mold	General	CTFA	18, 19
Mycological Agar (Mycophil)	Fungi	General	CTFA	20–22,
				31, 32
Mycophil Agar pH4.7	Fungi	General	CTFA, ASTM	20–22,
	Acidic bacteria			31, 32
	Yeast (saprophytic)			
Letheen Agar	Bacteria	Neutralize	CTFA, ASTM	13, 23, 25,
	Yeast/fungi			29
	Anaerob. microb.	General	CTFA	28
Tryptic Digest of Casein and Soy Agar	Yeast/mold	General	CTFA	26
Infusion Agars	Bacteria	General	CTFA	13
Thioglycollate Agar	Bacteria	Neutralize	CTFA	13
Tryptone-Glucose-Yeast Extract Agar	Bacteria	General	CTFA	18, 19
Trypticase-Glucose-Yeast Extract Agar	Bacteria	General	CTFA	18, 19
D-E Agar		Neutralize	CTFA	27, 46
Trypticase Soy Agar with lecithin		Neutralize	CTFA	BBL
and Polysorbate 80				
Soybean-Casein Digest Agar	Bacteria	General	USP	17, 26
	Yeast/fungi			

Broth				
Letheen Broth with lecithin	Bacteria Yeast/fungi	Neutralize Also to dilute	CTFA	13, 25
Letheen Broth with lecithin and Triton X-100	Bacteria Yeast/fungi	Neutralize Also to dilute	CTFA	13
Thioglycollate Broth	Bacteria Yeast/fungi	Neutralize Also to dilute	CTFA	13, 25
Williamson Buffered Suspending Fluid (Modified)	Bacteria Yeast/fungi	Neutralize Also to dilute	CTFA	13, 25

contact with the consumer; challenge tests do not measure the protection afforded by these delivery mechanisms.[24]

Once the sample is inoculated, the technician mixes the contents thoroughly. A sample is taken and diluted in the proper neutralization broth to inactivate the preservative. If the neutralizer's effectiveness can not be established, then physical dilution or membrane filtration may aid in inactivating the preservatives (see next section).

Most procedures require pour plating of 1 gram or 1 ml of the inoculated product. Some labs use an alternate streaking method to estimate the microorganisms present. Additionally, one may use spread plating, especially when the organisms are sensitive to temperatures required for tempering agar (45 to 47°C). The inoculated samples are incubated at room temperature or at a temperature that encourages proliferation of the test organisms used in the challenge. The incubation temperature for the plates is typically 32 to 37°C for bacteria and 25 to 30°C for fungi.

Most test method development groups (e.g., CTFA, USP, ASTM) recommend sampling on the following days after each challenge: 0, 1, 2, 3 (eye cosmetics only), 7, 14, and 28. Some tests may need more than 28 days, depending on its intended usage. Rechallenges are often done to estimate preservative adequacy in some products. Repeatedly challenging the product with a particular organism will show the number of challenges needed to inactivate the preservative system.[13]

The CTFA recommends diluting the inoculated sample with the following neutralizers: Letheen Broth with Lecithin, Letheen Broth with Lecithin and Triton X-100, Thioglycollate Broth, or Williamson Buffered Suspending Fluid.[13,25]

Lecithin or polysorbate 80 added to media is usually enough to neutralize most preservative carry-over and disperse the product when using the pour plate method.[13] Letheen Agar is a standard recommendation for bacteria, yeast, and fungi.[23] This medium was originally intended as an anaerobic growth medium.[28] It is effective at neutralizing quaternary ammonium compounds.[29] Other media recommended by the CTFA[13,23] include: Tryptic Digest of Casein and Soy Agar,[26] Nutrient Agar,[14] Thioglycollate Agar, Infusion Agar (brain, heart, veal, or combinations of), Eugonagar,[16] Tryptone-Glucose-Yeast Extract Agar or Trypticase-Glucose-Yeast Extract Agar,[18,19] D-E Medium,[27,47] and Trypticase Soy Agar with lecithin and Polysorbate 80 (BBL). Other media that may produce a more luxuriant growth of fungi include Saboraud Dextrose Agar,[17] Mycophil Agar and Mycological Agar,[20–22,31,32] and Eugonagar.[16]

A neutralizer should be incorporated in the plated agar when recovering bacteria by streaking plates.[34,35] Most of these methods recommend Letheen Agar.[13,25] If the preservative is a mercurial or other heavy metal, then one should use an agar with thioglycollate.

ASTM Method[36]

This method is used to test preserved samples compared to nonpreserved samples. The ASTM provides a list of challenge organisms (Table 4-1). The method recommends maintaining the microorganisms on nutrient agar[14] for bacteria and on Mycophil agar at pH 4.7 for fungi, yeast, and acidic bacteria.[20–22,31,32] Transfers should occur monthly with bacteria incubated at 32°C and fungi at 25°C. One prepares fresh cultures for the inoculum. This is done by growing them on the right solid medium for 18 to 24 hours at 37°C (bacteria) or 48 hours at 25°C (yeast). Fungal cultures should grow for 7 to 14 days (until full sporulation) at 25°C on the proper medium.

The method provides for harvesting the organisms with a sterile inoculating loop, and transferring it to sterile distilled water. The optical density is measured at 425 nm to yield 1.0×10^8 bacteria/ml. Fungal cultures and spores are dislodged by rubbing gently with a sterile inoculating loop or removing it with a sterile glass hockey stick. Then these are filtered with sterile nonabsorbent cotton to remove the hyphae and break up any clumps. One may use a hemocytometer count to adjust the spore level to 1.0×10^7/ml.

ASTM allows two types of challenges: a mixed culture method or a single culture method. The mixed culture challenge permits three separate inocula preparations. These preparations usually include equal portions of (1) Gram-positive bacteria, (2) Gram-negative bacteria, and (3) yeast and fungal suspensions. To determine the number of colony-forming units, the method uses serial dilution. Plating is done in duplicate using Letheen agar.[13,25] Incubation is done at 32°C for 24 hours for bacteria and yeasts; for fungi, incubation is done at 25°C for 72 hours.

ASTM suggests preparing the sample in three 100 gram samples in glass containers with lids. The method calls for inoculating each sample with 1 ml of each microorganism suspension (final concentration of 1×10^6 bacteria/ml or 1×10^5 yeasts or spores/ml). One then mixes these inoculated samples and stores them at ambient temperature. At the proper times (0, 7, 14, 21, and 28 days), 1 part test sample is mixed with 9 parts Letheen broth. Additional 10-fold dilutions are done and duplicate plating of each dilution is done with Letheen agar. Incubations of bacteria and yeast are done at 32°C (fungi at 25°C) for at least 72 hours. One counts the CFU per plate to determine the number of surviving microorganisms per gram of test product.

Some cosmetics receive repeated exposure or contamination. Where this is possible, ASTM allows a rechallenge with the microorganisms at 28 days. The ASTM, however, does not specify an inoculum level. The test would then continue for an additional 28 days.

The ASTM criteria require that Gram-positive and Gram-negative bacteria and yeasts should show at least a 99.9% decrease within 7 days following each challenge. There should be no increase after that for the remainder of the test. Within 28 days, fungi should show a 90% decrease and again show no increase within the test period. Unpreserved controls should fail both these criteria.

The ASTM procedure provides a method to check for neutralization of the preservative if no growth is seen on any plates. This is done by streaking plates from the 10^{-1} and 10^{-2} plates with a 10^{-3} dilution of nutrient broth culture of mixed inoculum. This mixed inoculum may be either gram-negative bacteria, gram-positive bacteria and yeast, or a mixed culture of fungi incubated 18 to 24 hours. Lack of growth after incubation at 32°C and 25°C for 72 hours suggests that neutralization of the biocide did not take place.

Unfortunately, this procedure does not prove that neutralization occurred at the time of plating. The original plates used for the test have already been incubated at least 3 days. Therefore, the preservative is not in the condition it was in at the time of sampling. Growth in this system cannot be taken as evidence of effective biocide neutralization at the time of the sampling. Another method offered by ASTM is detailed below which is far more appropriate.

USP XXII Method[15]

Table 4-1 lists the recommended test organisms. One may also include any other organisms that are likely contaminants. A medium such as Soybean-Casein Digest Agar supports vigorous growth. USP recommends it for initial cultivation of these organisms.[15,26]

Freshly grown stocks of a particular culture are prepared by inoculating a solid agar medium. Incubate bacterial cultures at 30 to 35°C for 18 to 24 hours. Incubate yeasts at 20 to 25°C for 48 hours, and fungi at 20° to 25°C for 1 week. One then harvests the bacteria and yeast using sterile saline (0.9% NaCl) and then dilutes them to 1×10^8 CFU/ml. Harvest fungi with sterile saline containing 0.05% polysorbate 80 and adjust the spore count to 1×10^8 CFU/ml. The number of colony forming units per ml determines the amount of inoculum to use in the test. The viability of the suspension should be monitored, especially if not used promptly.

A 20 ml sample of the product is transferred to a sterile, capped bacteriological tube if one cannot inoculate the product container and sample it aseptically. Inoculation of the test sample with the suspension is done using a ratio of 0.01 ml inoculum to 20 ml test sample. The concentration of microorganisms in solution should be between 1×10^5 and 1×10^6 CFU/ml. The viable number of microorganisms of the inoculum suspension is determined by the plate-count method. Use this value to calculate the initial concentration of CFU/ml in the test product.

The inoculated containers are incubated at 20 to 25°C and examined at 7, 14, 21, and 28 days after inoculation. The number of CFU/ml are determined by the plate-count method at each of these intervals. Then, one can calculate percentage change from the initial viability.

According to this method, an effectively preserved system will reduce the viable microorganisms to less than 0.1% of the initial concentration by the 14th day for bacteria. For yeasts and molds, the viable level must decrease or remain the same. The concentration for all microorganisms must remain at or below these designated levels for the remainder of the test.

Comparison of Methods

The ASTM gives the most detail of the three methods. The CFTA leaves more room for customizing a test for a specific target. The USP recognizes that many drugs are not subjected to the same rigors of consumer use and abuse as are cosmetics. Table 4-2 provides an overview of the methods.

Challenge Microorganisms

Challenging the product with appropriate organisms is a major concern in determining how effective a preservative needs to be. Organisms representing possible contaminants either through consumer use or manufacture are ideal for this purpose. Since organisms can develop resistance to preservatives and cause opportunistic infections,[1,4] we must always be looking for these new sources of challenge organisms.

Staphylococcus aureus (ATCC 6538) is a common skin organism.[14] Most preservative challenge test methods use it to challenge frequently used cosmetic products since it is a common contaminant that may pose a threat to consumers.[4] It represents the Gram positive cocci in many tests. Since it requires comparatively demanding nutrient needs, it does not always seem to be such a logical choice for a challenge inoculum.

Pseudomonas aeruginosa is a nonfermentative Gram negative rod suggested by all three methods. The ASTM and USP recommend ATCC strains 9027 and the CTFA recommends strains 15442 and 13388. *P. aeruginosa* is a well known, highly ubiquitous pathogen. It also shows high resistance to many preservatives.[23]

Both the CTFA and the USP methods recommend *Escherichia coli*, ATCC 8739, a fermentative Gram negative rod. It is a member of the largest bacterial families, the Enterobacteriaceae. It is an indicator organism of fecal contamination.[23] Like most of the coliforms, it can easily develop biocide resistance. *Enterobacter (Aerobacter), Klebsiella* and the *Proteus* spp. are sometimes substituted for *E. coli.* ASTM even recommends *Enterobacter aerogenes* (ATCC 13048) instead of *E. coli.*

All three methods recommend *Candida albicans* as the naturally occurring, representative yeast. It can be pathogenic[23] and can represent the resistance of yeasts to preserved systems. The ASTM and USP recommend ATCC strains 10231 while the CTFA does not recommend a specific ATCC strain.

A major cause of product decomposition is contamination by filamentous fungi such as *Penicillium* or *Aspergillus* spp.[23] All three methods recommend *Aspergillus niger.* The ASTM and USP use ATCC strains 16404, while the CTFA uses strain 9642. In addition, the CTFA also suggests *Penicillium luteum* ATCC 9644.

Microorganisms that are indigenous to the normal eye, clinically significant isolates, and product isolates, are what CTFA recommends when challenging eye cosmetics. These include the organisms detailed in Table 4-1. Gram positive spore formers are represented by *Bacillus subtilis.* Eye cosmetics or any inadequately preserved systems may allow Gram- positive spore formers to survive, germinate, and actively proliferate. Using a Gram-positive spore former, such as *B. subtilis,* for a challenge inoculum should be done carefully in order to show the susceptibility of the vegetative forms to the preservative. If the inoculum preparation procedure promotes sporulation, then the challenge may be too rigorous for a product to pass until it is over-preserved and unsafe to use toxicologically.

Maintenance and Harvesting of Organisms

The ASTM and USP give specific recommendations for maintenance of the microorganisms. The CTFA specifies the media but only recommends periodic subculturing. This is an important consideration as standardization of culture preparation is a critical concern for reproducible results. Considerable work has been done by Peter Gilbert and his lab to show that a great deal of phenotypic variance occurs simply due to growing organisms under nutrient -limited conditions such as occur when grown to late log/early stationary phase.[12,12a] The ASTM and USP methods also include harvesting and standardizing conditions. This may involve filtration of fungal spores.[36]

Harvesting methodology differs between the ASTM and USP methods. This may influence the viability of the organisms.[36] Neither method specifies buffering the solutions to the pH or ionic range of the culture medium. Suspension of the cells in a solution that is at an incorrect pH or osmolarity than the culture medium may have inhibitory toxic effects.[37] Orth found[38] that broth inocula decreased the rate of inactivation of the test organisms compared to the use of saline inocula prepared from surface growth on agar media. Although not specifically mentioned by Orth, this result was likely due to the broth medium serving as a preservative inactivating agent.

Preparation and Standardization of Inoculum

Dilution of the cell suspension should not occur in an unfavorable environment. Diluted cells are more susceptible to being harmed than denser cell concentrations. A buffered solution will protect against a pH change after the cells have been suspended.[37] The ASTM uses sterile water while the USP uses sterile saline when washing the organisms from the transferred stock culture. The CTFA does not specify any recommendations for harvesting.

The ASTM is specific as to standardization of the bacteria, yeast, and fungi. It recommends a certain spectrophotometer, the spectrophotometer tubes, and the absorbance wavelength. Neither the USP or CFTA are as specific. They require only a final inoculum level. The inoculum level is different among the three methods. Before inoculation, the ASTM recommends that the concentration of the microbial suspension be 1×10^8 CFU/ml for bacteria and 1×10^7 CFU/ml for yeasts and fungi. The USP recommends adjusting the microbial or spore count to 1×10^8 CFU/ml while the CTFA does not address standardization of the suspension itself.

The resulting challenge level of the product is more similar between the methods than the inoculum levels. All three methods recommend a level of 1×10^6 CFU/ml or gram of product for bacteria. The USP actually gives a range between 10^5 and 10^6. For yeast and fungi, the CFU levels per ml or gram are slightly different. The ASTM recommends 1×10^5, the CTFA 1×10^4 and the USP again gives a range of 1×10^5 to 1×10^6. The CTFA challenge level is comparatively low. As a result, this level limits the measurable reduction to less than 4 log units.

Pure vs. Mixed Cultures

The ASTM recommends using a pure or mixed culture inoculum, while the CTFA and USP are typically pure culture tests. Mixed cultures may more accurately reflect the normal contamination profile of a product used by a consumer. However, pure cultures may exhibit more resistance to the preservative than a mixed culture.[4,39] One recommendation is that organisms should be related when using mixed cultures.[4] For example, one would combine Gram-positive species or Gram-negative species.

Incubation Conditions, Interpretation, Rechallenge

All three methods state sample and inoculum size, while the ASTM and USP also define storage conditions for the samples. All three methods require sampling after inoculation at 0, 7, 14, and 28 days. The CTFA also suggests sampling at day 1 and 2, with a 3 day sampling for eye cosmetics. By 7 days, according to the CTFA and ASTM, the vegetative bacterial counts should reach a 3 log or 99.9% reduction. The USP allows 14 days for this reduction of bacteria. The yeast and fungal counts must be reduced by 90.0% at the 7th day for the CTFA and 28 days for the ASTM. The USP only specifies that by 14 days the level of yeasts and fungi remain the same or below the initial concentration within a certain tolerance level (e.g., 0.5 log).

Only the ASTM recommends a rechallenge with microorganisms at 28 days. Repeat challenges may provide a better indication of potential problems of product contamination while in consumer use.[1] For example, this may be important for assessing hand and body lotion contamination or mascaras which are repeatedly used. Rechallenging or repeating the inoculations may indicate how the preservative system of a particular product would withstand insult before it failed.[24]

Other Published Methods

D-value methods

Orth proposed a rapid method for estimating preservative efficacy.[38,38a] This method uses short sample times and estimates the final response at 28 days by linear regression as a D-value. Orth claims that each organism has a characteristic rate of death. When this rate is multiplied by the log of the inoculum challenge, it can predict the time it takes to inactivate the entire challenge.

This method is not a new development, but an adaptation from food microbiology heat destruction D-values. One weakness inherent in this method is the extrapolation of kill beyond the measured data. This is not valid for linear regression analysis.[40]

A second weakness of the method is that it assumes a linear relationship between time of exposure to the biocide and number of surviving microorganisms. One can usually handle the logarithmic nature of biocide killing by log transforms. However, even this relationship does not exist for a variety of microorganisms and biocides.[41]

If one performs all the D-value assays precisely the same, it may be possible to show reproducible results using this method. This would allow a rapid preliminary screening of preservatives, but should not be relied upon as the sole method of testing.

Capacity Tests

The capacity test assesses the efficacy of concentration and antimicrobial spectrum of a preserved cream, suspension, or solution. In this assay, mixed bacterial cultures are grown in nutrient broth for 48 hours. Yeasts and molds are grown in 2 per cent v/v Malt Extract in distilled water.[42] The test method permits storage of mixed mold spore suspensions in distilled water at 4°C. These suspensions are from cultures grown on Malt Extract Agar plates.

In a capacity study by Barnes and Denton,[42] the mixed cultures consisted of: (1) Gram-negative bacteria (*Escherichia coli, Proteus vulgaris, Pseudomonas aeruginosa, P. fluorescens*), (2) Gram-positive bacteria (*Staphylococcus aureus, S. albus, Micrococcus flavus, Sarcina lutea*), (3) aerobic spore formers (*Bacillus subtilis, Bacillus cereus, Bacillus megaterium*), (4) mold spores (*Mucor plumbeus, Aspergillus niger, Cladosporium herbarum, Penicillium spinulosum, Trichoderma* spp.), (5) and yeasts (*Saccharomyces cerevisiae, Sporobolomyces* spp., *Schizosaccharomyces pombe, Candida albicans*).

In the study, the preservatives tested included benzalkonium chloride, Bronopol, chlorhexidine gluconate, chlorocresol, Dowicil 200, methyl parabens, Phenonip, propyl parabens, thimerosal, and a "Preservative C." Barnes and Denton incorporated the preservatives into creams, suspensions, or solutions at a recommended use level. They also tested two lower concentrations.

The test protocol required a thorough mixing of 1 ml inoculum into 20 grams of the cream, suspension, or test formula. They stored the formulae at room temperature for 48 hours. They sampled the creams and suspensions with a sterile loop while the solutions had 1 ml removed. They dispersed this sample into either 5 ml (for the creams and suspensions) or 9 ml (for the solutions) of nutrient broth with Lubrol® and lecithin as neutralizers. This applied for all preservatives except thimerosal, which required sodium thioglycolate as the neutralizer.

A sample of this dispersion was then plated on nutrient agar (bacteria) or Malt Extract (molds and yeasts) containing neutralizers. They incubated the plates at 37°C for 24 hours (bacteria) or for 25°C for 48 hours (molds and yeasts). The authors repeatedly cycled through the reinoculation and sampling for 15 times or until three consecutive positive results occurred.

A preservative must reduce the number of viable organisms in the inoculated formulation by 10^3 in 48 h for creams and suspensions to produce a single negative result according to Barnes and Denton.[42] This ability diminishes gradually due to dilution and biocide absorption by the added organisms.

H.S. Bean[9] also cited the need for a "performance test" rather than the typical preservative efficacy test. He felt that such a test would measure "the ability of the contaminating organisms to destroy the product." Such a test may be more rapid than a typical preservative efficacy test and more quantitative in assessing the ability of a product to handle contamination.

Predictive Tests of Consumer Contamination

Brannan et al.[43] conducted a study designed to validate the CTFA preservative efficacy test as a predictive model of consumer contamination. This is a critical concern. One can control the microbial insults from the manufacturer by sanitary processing. However, the product must also provide consumer protection from pathogens during use. Brannan et al.[42] evaluated two cosmetic formulations, a lotion, and a shampoo at three different preservative levels. They used a modified CTFA preservative efficacy test method. They first challenged products diluted to four concentrations (30, 50, 70, and 100%) with mixed cultures of bacteria known to contaminate cosmetics. After 28 days, they used these results to classify the formulations as either poorly preserved, marginally preserved, or well preserved. The formulae were then used by consumers and evaluated after use for contamination.

For the consumer contamination part of the study, they packed the products in containers to allow direct contact with the consumer. This assured that package design was not a significant factor in preventing exposure to microbial contaminants from the consumer or the environment. They defined contamination as recovery of >100 CFU/g or if Gram-negative bacteria were present at initial receipt and 4 to 7 days post-receipt.

The well preserved products had no samples contaminated after consumer use, while the poorly preserved had 46 to 90% of the products contaminated after consumer use. The marginally preserved products had 0 to 21% contaminated. Thus, the method accurately predicted the potential for consumer contamination. It did not, however, account for the significant role that the container plays in preventing consumer contamination.

General Considerations in Formulating Preserved Products

Interactions—Oil-based Emulsions

A preservative may have its chemical and biological activity influenced by the overall formulation of the product. For example, a minor change in pH may change the ionic character of the preservative, change the chemical groupings on the bacterial surface, or increase the partitioning of the preservative between the product and microbial cell.[9]

Some preservatives can bind to surfactants. Nonionic surfactants in particular can impair the antimicrobial activity of some preservatives, such as parabens. Other preservatives will be inactivated by proteins and exhibit reduced antimicrobial activity as a result.[44] In oil-based emulsions, most of the emulgent disperses throughout the water phase as emulgent micelles. This redistributes the preservative and changes the concentration level in the aqueous phase.[9] Additionally, the buffer system may affect the activity of the preservative. For example, borate buffered formulations are more easily preserved than phosphate buffered formulations, presumably because the boric acid also acts as an inhibitor of membrane potential. One must understand all factors of the formulation, not just the ingredients recognized as the preservative.

The level of solids present in a formulation can also impact the effectiveness of the preservative. Inorganic solids, (carbonates, silicates, or oxides) and organic solids (cellulose or starch) absorb preservatives such that one must use higher concentrations. Talc, for example, decreases the antimicrobial activity of methyl paraben by as much as 90%.[44]

Container Considerations

The type of container used for packaging a cosmetic will influence the concentration and activity of the preservative.[24] Generally, the more lipid-soluble preservatives are absorbed by containers and their closures.[9] One must test containers to determine the actual preservative effectiveness under actual storage conditions. Adsorption, complexation, or volatility can erode antimicrobial activity. Certain containers are not compatible with some preservatives, such as nylon and parabens or polyethylene with certain phenolics, mercurials, and benzoates.[44]

Dispensing closures can also be important considerations in preventing microbial contamination during consumer use. A study by Brannan and Dille[45] showed that, during consumer use, unpreserved shampoo with a flip-cap had the greatest degree of protection from contamination

(0%). For an unpreserved skin lotion, a pump-top dispenser afforded the best protection from contamination (10%). Other types of closures tested included the standard screw-cap and slit-cap. The screw-cap closure provided the least amount of protection, while the slit-cap provided moderate protection from contamination. This study underscores the need for considering preservation as an attribute of the entire product, not just the active preservative. This study also points out the fallacy of logic that PET can *ever* be predictive of consumer contamination potential.

NEUTRALIZER EVALUATION AND MICROBIAL CONTENT TESTING

Function of Neutralizers

Neutralizers should inactivate the preservative or biocide and allow for unrestrained microbial growth.[48] If the biocide is not inactivated, one will overestimate the antimicrobial activity of the biocidal agent since killing will continue in the recovery medium.[47] Included in the evaluation of the neutralizer should be controls to measure neutralizer toxicity toward the microorganisms.

Types of Neutralization

Chemical

Lecithin, polysorbate 80, and sodium thiosulfate are examples of chemical inactivators for quaternary ammonium compounds, phenolics and halogens, respectively. The type of neutralizer and the effective concentration need to be determined for each biocide and microorganism permutation (see Table 4-4). Disinfectants and preserved products are typically tested by inoculating them either directly or via microbially contaminated carriers. When sampling the inoculated disinfectant or preserved product, one must stop the killing activity of the biocide immediately as the disinfectant or product is sampled.

Some biocides are difficult to inactivate. For example, the neutralizer for formaldehyde and glutaraldehyde is sodium bisulfite. Unfortunately, it will also inhibit growth of bacteria and germination of spores.[11] Thus, it is equally important that the neutralizer be nontoxic to the microorganisms. One determines the neutralizer toxicity by comparing growth in the neutralizing medium alone to growth in a typical medium such as Tryptic Soy Agar or Sabouraud Dextrose Agar.[30] Most test procedures require the chemical neutralizers to be included as part of the dilution broths into which one places the sample. One may also include the chemical neutralizer in the plating agar.

An effective general neutralizing medium is Dey-Engley. It contains the neutralizing agents sodium thioglycollate, sodium thiosulfate, sodium bisulfite, lecithin, and polysorbate 80.[27,30] It is available in a broth or agar preparation. Some common diluting fluids are in the USP XXII: DFA with

TABLE 4-4

NEUTRALIZING AGENTS IN DISINFECTANT TESTING[46]

Substance (or group)	Neutralizing Agent or Dilution
Aldehydes	Dilution Sodium Sulphite Glycine Dimedone-morpholine?
Phenolics	Dilution Polysorbate 80
Mercury compounds	Sodium thioglycollate, Cysteine
Hydrogen peroxide	Catalase
Alcohols	Dilution
Organic acids and esters	Dilution Polysorbate 20 or 80
Acridines	Nucleic acids
Quaternary ammonium compounds (QACs), biguanides	Lecithin + Lubrol W Lecithin + Polysorbate 80
Tego compounds	Polysorbate 80
EDTA and related chelating agents	Dilution Mg^{2+}
Hypochlorites	Sodium thiosulphate Nutrient broth
Iodine	Polysorbate 80 Sodium thiosulphate

0.1% meat peptone, DFD with meat peptone plus polysorbate 80, and DFK with meat peptone, polysorbate 80, and beef extract. Recently, Sutton et al.[48] developed a Universal Diluting Fluid that inactivates a broad range of commonly used biocides for membrane sterility testing in eye care products.

Dilution and Membrane Filtration

Some preservatives are sensitive to concentration and are effectively neutralized by dilution (see Table 4-4). There are several excellent papers on this subject.[11,33,47] Another method of neutralizing a preservative is membrane filtration. The bacteria are isolated on the filter and then rinsed free of preservative. One then transfers the surviving bacteria to a growth medium. Bloomfield[11] suggests washing the membrane with nutrient broth and transferring it to the surface of an agar plate to count surviving colonies. The nutrients from the agar diffuse through the membrane to support the microorganism's ability to grow into visible colonies.

One can test for neutralization of the biocide on the membrane filter by placing a known amount of microorganisms on the membrane without the biocide. One then places the same amount of microorganisms on another membrane that has been exposed to disinfectant. The disinfectant is rinsed through (and thus off) the membrane filter by placing neutralizing agent onto the filter and allowing the neutralizer to go through the filter either using a vacuum or pressure. If the two membranes show equal counts, then the disinfectant has been neutralized. If the disinfectant is bound to the membrane, then counts will be significantly less for that membrane compared to the one not exposed to the biocide.[11]

ASTM Methods for Testing Biocide Neutralizers[49]

The ASTM methods are actually a series of experiments to show a neutralizer is nontoxic and effective. Testing is first done to determine Maximum Tolerated Concentration of inactivator (MTC). This involves using the right liquid test media. Often, this will be buffered peptone water. After adding the target organism, one determines the microbial counts at zero and after 30 minutes. The peptone water should not cause a significant decrease in microbial survival over the 30 minutes.

One can add several concentrations of the neutralizer to the peptone water system. Then, one compares the microbial survivors of the 30 minute exposure in neutralizer/peptone solution to the 30 minute peptone control survivors. This comparison gives the maximum tolerated concentration (MTC) of neutralizer that does not decrease the microbial survivors of the targeted microorganism.

The next procedure determines the effect of the MTC on the antimicrobial. One adds specific concentrations of the antimicrobial to the MTC of neutralizer and compares the numbers of microbial survivors at the 30 minute time point. The survivors of the neutralizer-antimicrobial system should not differ significantly from the counts of the peptone/neutralizer controls.

The ASTM method uses square-root transforms of the plate count data for the different treatment groups and controls. It assumes that the plate count data are Poisson-distributed data. The method then uses the t test to test for significant mean differences.

Microbial Content Testing

With the knowledge that microbes can grow in cosmetics comes the responsibility for detecting them. The currently used methods include traditional plate counting and enrichment testing. Microbiologists are rediscovering that plate count procedures are, in fact, invalid predictors of the number of viable organisms present in a sample.[50] Couple this flawed estimate of bacterial numbers with high serial dilution errors[51] and it is surprising that we continue using such archaic and inaccurate methods. We had data over 50 years ago that told us not to use such methods, yet we still cling to them. Perhaps the problem is that not only are they simple and inexpensive but they seem to work quite well in a clinical setting where results are measured in terms of life and death. Plate counts are, however, absolutely inaccurate for anything more than a very rough estimate of the numbers of microbes present. Couple this with the fact that most media are designed to detect clinical isolates and we have even more reason to abandon a method that is incapable of correctly predicting the numbers of microbes in an environmental sample whether that sample is water or a consumer product.

The plate count is a pragmatic and practical method that has survived because of its simplicity, not because of its scientific validity. Thus, we must continue using it until new methods supersede it. Some of the rapid methods offer hope, but it is surprising that they are measured against the flawed standard of the plate count method and serial dilutions when they are evaluated. They should instead be evaluated against a microscopic count that considers numbers as well as clumps that give rise to colonies. It would be a fascinating study to reinterpret the data used to support the concept of injured organisms with a new paradigm that bacteria under stress clump and only those on the outside of the clump get killed.

An emphasis must be placed on the concept that the aerobic plate count (APC) in cosmetic microbiology is strictly an estimate of the numbers of viable aerobic organisms in the product. Data should always be presented in terms of colony forming units (CFU) and the dogma that one colony arises from a single organism must be rejected before clearly understanding the limitations of a plate count.[52]

Considerable information is devoted to a plate counting method's ability to prevent "preservative carryover" (e.g., to neutralize all biocides present) and to its being able to allow "stressed organisms" to recover and grow into a colony forming unit.[53] Most commonly used are a combination of Letheen broth as the initial diluent for the product sample and Tryptic Soy Agar with Lecithin and Tween 80 for the plating medium into which the diluted sample is placed (Orth). Many other neutralizers can be used as well.[27,29,30,46–49] Test methods for determining neutralization efficacy are also well described.[30,47,49] Retroactive test methods that are commonly used (ASTM and AOAC) are entirely inappropriate since they only show that neutralization finally occurred at the time of testing for neutralization and not at the time of testing the sample for microbial content.[52]

Several sources may be referred to for additional information on methods for plate counting of consumer products to determine microbial counts.[5,23,26]

Product and Raw Materials Tests

In these tests, the microbiologist is trying to obtain an estimate of the aerobic population of potential contaminants in the product itself. The product is sampled aseptically (usually in 1 gram amounts) into a diluent meant to both neutralize the product's antimicrobial nature as well as dilute the organisms present to countable values. The diluted and neutralized product is then plated into a medium that promotes growth and recovery of injured organisms.[54]

In order to conduct any of the content tests, a well supplied microbiology lab is needed. Many lists of the needed equipment and supplies exist (FDA manual, USP, CTFA) and will not be enumerated here. Each lab should document in their SOPs how the equipment should be used, how the materials and reagents should be made (e.g., media, dyes, biochemical test solutions), and how the tests should be conducted (e.g., aseptic technique defined, room air circulation, and handling defined).

Most products and raw materials are received as they are packed for consumer use or in sterile containers. Typically, if either the product or raw material is aqueous, then 10 grams of product or raw material is placed into 90 ml of a dilution blank with appropriate neutralizing media. The bottle is then capped and shaken thoroughly. Anywhere from 0.5 to 1.0 ml of this dilution is pour plated or 0.1 ml is spread plated on the recovery medium. The plates are incubated at 33 to 37°C for anywhere from a few days all the way to a week if lower temperatures are employed for detection of yeasts and mold. Some microbiologists prefer using a specific medium for recovery of yeasts and mold (e.g., potato extract agar or Sabouraud). The colonies on the plate are counted and the value is multiplied by the dilution factors required.

If the product is a powder or a solid, then it must be dissolved or resuspended into the diluent using solubilizing polysorbate 80 or slight heat before plating.

Package Tests

These tests are usually conducted on an empty package as received from the supplier. They are not expected to be free of microorganisms. Typically, one pours a specific amount of sterile diluent into the bottle and it is capped with the cap used in ordinary production. The bottle is then shaken thoroughly and the diluent is plated or diluted further with sterile diluent and plated. The plates are incubated and counted as described previously. The colony forming units (CFU) per sample are determined by multiplying the count by the appropriate dilution factors employed in the test.

Environmental Tests and Monitoring

A variety of environmental tests may be conducted to determine bioburden in the air or on the equipment. Typically, in cosmetic manufacture, the only concern is for the number of organisms that are on the product contact surfaces in the plant. A simple way to conduct these tests is to use a swab moistened with Letheen broth (or any other neutralizing diluent that would support growth) and swab the equipment. The portion of the swab that is not contaminated by technician handling is broken off into the medium and incubated. If turbidity occurs after incubation, then sterility had not been achieved on the product contact areas. Alternatively, the diluent may be immediately plated or diluted further and plated to determine the actual numbers of contaminants per surface area.

Identification of Microbes
Considerable space is typically devoted in the literature to identification of the microbes isolated from the above tests.[5,55] A thorough resource for identification of microorganisms can be found in the ASM's Manual for Clinical Microbiology.[56] Use of rapid identification techniques is replacing many of the traditional culture methods. Although these can be adapted from their clinical uses to cosmetic microbiology, it must be kept in mind that most of the organisms encountered in cosmetic microbiology are environmental isolates. Thus, rapid ID systems geared for environmental isolates should be preferred over those developed primarily for clinical use.

DISINFECTANT TEST METHODS

Disinfection is the removal or destruction of pathogenic microorganisms from inanimate objects or surfaces, usually by use of a chemical agent. For disinfectants to be of practical use, they must have a rapid, lethal antimicrobial effect. Block cites five elements in the definition of a disinfectant: (1) removes infection, (2) kills, not just inhibits, microorganisms in the vegetative stage, (3) does not need to kill spores, (4) is ordinarily a chemical but can be a physical agent, and (5) is used only on inanimate objects, not on the human or animal body.[6]

The bacteria used in disinfectant testing should represent the main bacterial groups: bacilli and cocci, Gram-positive and Gram-negative, and mycobacteria in certain standards. Use of appropriate yeast and fungal inocula is also recommended, especially if exposure of the product to these microorganisms is likely. Use of other microorganisms that also might be present is appropriate. The inoculum should show a minimum decrease of 10^5 cells/ml for all microorganisms to show a significant statistical value.[6]

D-Value Methods

When subjected to lethal treatment, every organism should have its own characteristic rate of death.[57] One means of measuring disinfection efficacy is by D-values, the decimal reduction time or rate of inactivation of microbes by a disinfectant. More specifically, this rate is the amount of time required for the disinfectant to inactivate 90% of the viable microbial population. It can also be thought of as the time to decrease a microbial population by 1 logarithmic unit. If placed on a graph, with log numbers of microorganisms on the y-axis and time on the x-axis, then the D-value is the negative reciprocal of this slope.[6]

Microbiologists usually find this value by plotting the log of the number of microorganism survivors per milliliter or gram vs. time after initial inoculation. By linear regression, they estimate the X-intercept. This intercept describes the activity of a disinfectant by a single number, the

D-value. This identifies the time required for the number of surviving organisms to decrease to 1 CFU/ml (i.e., where the log 1 CFU = 0).[6] The goal of determining the D-value is to get a single and quantitative indicator of the rate of microorganism death.

There are three different ways to determine D-values. Each may estimate a different D-value particularly if the rate of microorganism death does not follow first order of kinetics.[41] The most commonly used method is the Stumbo method or end-point method using only initial and terminal points. In this method, one exposes a known level of inoculum to the disinfectant for a specified amount of time. One then determines the number of survivors. The D-value is the ratio of time it takes for a $\log_{10}$ reduction in CFU per milliliter. The second method involves averaging multiple end-point D-values, one for each sampling, derived over the course of an experiment. The third approach uses linear regression analysis of multiple data points.

The D-value grew out of the food industry and the use of thermal disinfection. When using heat, the relationship between log survivors and time is linear. However, one should not assume a linear relationship for chemical biocides.[41] If the kill kinetics of the microorganism are nonlinear, one will get different D-values, depending on the method used for calculation. When kill kinetics are not first order, as in the case of many biocides, the D-value may falsely estimate the actual time required for disinfection.

Carrier and Suspension Methods

There are two methods of contact between the disinfectant and the microorganisms: carrier and suspension. The carrier method involves adding the inoculum to the carrier (such as watch glass, stainless steel cylinder, etc.) and allowing it to dry. One then places the carrier into the disinfectant. The controls must consider the spontaneous mortality of the microorganisms upon desiccation. The suspension method uses a liquid phase—the disinfectant diluted in distilled water.. The presence of organic matter, such as in peptone water, broth, or buffer, may interfere with the disinfectant activity. Neither method is without problems. The carrier technique tries to reproduce the surface of the area to be disinfected. The suspension method, however, can give valuable quantitative information on the disinfectant killing rate.[6]

To accurately measure the disinfectant activity, the microbiologist must eliminate any residual inhibitory effects from the disinfectant upon sampling and plating. This is done by neutralization when a sample is taken. Many of the neutralization methods are based on streaking plates with no growth and seeing if growth now occurs.[36] The major false assumption with this method is that proof of neutralization 3 to 5 days after sampling indicates neutralization at the time of sampling. It does not.

SKIN DEGERMING METHODS

Importance of Hand Washing

Medical staff play a significant role in transmitting infections as shown in many studies. As early as 1846, Semmelweis[58,59] dramatically lowered the maternal mortality rate from 13.7 to 1.3% in maternity wards. He did this by introducing a rigid regimen of hand disinfection with a solution of chlorinated lime. Today, all procedures involving direct or indirect patient contacts include hand hygiene or disinfection. The general feeling is that before surgery or invasive procedures, hand-washing should include an antiseptic. For routine patient contacts, hand-washing with soap and water is adequate.[60]

Ayliffe[61] showed that 70% alcohol is effective as a hand disinfectant for healthcare personnel. However, many organisms can survive this treatment. This observation led to a study where the investigators added a dye to the alcohol and observed hand-washing practices.[62] They found that the hand least used was often more thoroughly washed. Also, they found that people did not wash the thumbs as thoroughly. This appendage is more likely to come into contact with the patient or materials than any other finger.

This study shows the need for both an effective washing technique and an adequate disinfectant for effective hand disinfection. It also shows the need for teaching people how to wash their hands and to have a standard operating procedure for washing hands.

Types of Microbial Flora

In 1938, Price[63] divided the skin flora into two classes: residents and transients. Microorganisms that survive to multiply and colonize the skin are resident flora. Most skin antiseptics are somewhat ineffective against the resident flora. Resident flora are hardier (more resistant) and more persistent than transient flora. Different parts of the body of course have different flora, but most pervasive are the aerobic staphylococci (*Staphylococcus epidermidis*, *S. aureus*), micrococci, diphtheroids, and *Propionibacterium acnes*.[64,65,67,68] Typically the hair, face, axilla, and groin harbor the highest amounts of bacteria, while the arms and hands harbor lower levels. Of the hands, the area around and under the fingernails show the most microorganisms.

According to Noble and Sommerville,[68] other organisms may be present in small numbers. These include *Sarcina*, yeasts, and Gram-negative bacilli. *Staphylococcus aureus* colonizes intact, damaged, or abnormal skin. Only 20% of organisms are in the depths of the skin while the rest are near the surface.[69] Microflora located in the hair follicle are especially inaccessible.[60] Resident flora rarely cause infections unless the skin barrier is compromised. This may occur during an invasive procedure, such as surgery or catheterization. Newborn infants and patients with depressed host resistance are also susceptible.[70,71]

Transient flora, such as intestinal Gram-negative bacilli, do not survive or colonize normal skin well. Examples of transient bacteria include the *Streptococcus, E. coli* and *Pseudomonas* species.[53] One can remove the transient bacteria more easily from the skin than the residents. Most organisms survive poorly on the surface of the skin. This selection occurs because most organisms are poorly resistant to air drying and to the production of acidic by-products from other organisms on the skin.[74–82] Some organisms can be either resident or transient flora such as *Staphylococcus aureus* and various Gram-negative organisms.[74–82]

The goal of a preoperative scrub is to remove or kill as many transients and resident flora as possible. By binding to the skin, a disinfectant may continue exerting a residual effect for some time. This is important during surgery. Under glove occlusion, the skin flora multiplies to extremely high levels. If a nick in the glove occurs, the bacteria-laden sweat can leak into the patient.[83]

A perceived danger of repeated use of antiseptics is that they may destroy much of the normal resident flora and allow colonization with other microorganisms. This supposedly upsets the ecosystem of the skin and allows the skin's natural defenses to be ineffective against opportunistic bacteria. Although this hypothesis is intuitively logical, few studies have conclusively proven it.

Methods

ASTM Method for Evaluating Health Care Handwashes[84]

In the ASTM method, the degerming effectiveness of antimicrobial hand washing agents is compared through two sets of artificially contaminated hands. One group washes with the antimicrobial hand-wash and the other group is unwashed. The bacteria used to contaminate the hands is *Serratia marcescens*, a strain that has a pigment to help identification.

The volunteers refrain from using any antimicrobials for the duration of the test and at least 1 week before the test. This is often difficult as antimicrobials are found everywhere in antiperspirants, deodorants, shampoos, lotions, and soaps. In addition, the participants avoid materials such as acids, bases, and solvents as well as swimming pools, spas, and hot tubs.

The test is started by an initial 30 second practice wash to mimic the procedure. This was done to familiarize the panelists with the washing technique. Then the test panelists are inoculated with 5 ml of the *Serratia marcescens* suspension at 10 CFU per ml. Only the hands are exposed to the organisms. The hands are air dried away from the body for 1 minute. The researchers take a baseline bacterial sample by placing rubber gloves on the test hands and adding 75 mls of sampling solution to the glove. They then secure the gloves above the wrists and uniformly massage the hands for 1 minute. The technician then aseptically sample the fluid by standard microbiological techniques, such as membrane filter technique or surface inoculation.

The hand washing begins by the application of 5 ml (or a specified amount) of test formulation to the hands. The panelist then rubs the test material over all the surfaces of the hand and lower third of the forearm, completely lathering with a small amount of tap water for 30 seconds. The tap water is kept at 40 ± 2°C. The panelists repeat this washing procedure seven times. The researchers then inoculate the panelists with *S. marcescens* before the first, third, fifth, and seventh washes.

Sampling is done into Butterfield's sterile phosphate buffered water (pH 7.2) with a suitable inactivator for the antimicrobial. Plating is done with soybean-casein digest agar containing a suitable inactivator. The plates are incubated for 48 hours at 25 ± 2°C. Changes in CFU from the baseline counts until the sampling interval are determined by the reduction of the red pigmented *S. marcescens* plate counts. The reduction caused by a test formulation can then be compared to the control.

Rotter and Birmingham Methods

Other tests for hygienic hand disinfection involve similar procedures.[85] Two tests, the Rotter, or Vienna Test Model,[59] and the Birmingham, use *E. coli* as the test organism. In the Rotter procedure, volunteers immerse their hands in a broth culture. Afterwards, they allow their hands to air dry for 3 minutes. The disinfectant (3 ml of 60% isopropanol by volume) is then rubbed onto the hands for 30 seconds. Then, another 3 ml of disinfectant is rubbed onto the hands. To recover the organisms, the fingertips and thumb are rubbed against the base of a petri dish containing 10 ml of sampling fluid for 1 minute.

In the Birmingham test, *E. coli* is applied to the fingertips and rubbed until dry. The disinfectant is then applied for 30 seconds. The panelists then rub their fingertips and thumbs in a bowl containing 100 ml of sampling fluid and glass beads. Vigorous rubbing on the beads is done for 1 minute. Recovery of organisms is assayed both before and after disinfection for both tests.

REFERENCES

1. Durant, C. and P. Higdon. 1991. Methods for Assessing Antimicrobial Activity, In S. P. Denyer and W. B. Hugo, (Ed.), Society for Applied Bacteriology Technical Series no. 27, *Mechanisms of Action of Chemical Biocides*. Blackwell Scientific Publications.
2. Wilson, L. A. and D. G. Ahearn. 1977. Pseudomonas-induced Corneal Ulcers Associated with Contaminated Eye Mascaras. *Am. J. Ophthalmol.*, 84, 112.
3. Federal Register. October, 1977, 40, 53837-54838.
4. Cowen, R. A. and B. Steiger. 1976. Antimicrobial activity - a critical review of test methods of preservative efficacy. *J. Soc. Cosmet. Chem.*, 27, 467.
5. *FDA Bacteriological Analytical Manual*, 7th ed., 1992 Association of Official Analytical Chemists, Washington, D.C.

6. Block, S. S. 1992. Disinfection, Sterilization, and Preservation, Fourth Edition. pages 18, 887, 1009. Lea & Febiger, Philadelphia.
7. Tenenbaum, S. 1967. Pseudomonads in Cosmetics. *J. Soc. Cosmet. Chem.*, 18, 797.
8. Kabara, J. J. 1984. *Cosmetics and Drug Preservation, Principles and Practices.* Marcel Dekker, Inc., New York.
9. Bean, H. S. 1972. Preservatives for Pharmaceuticals. *J. Soc. Cosmet. Chem.*, 23, 703.
10. Chapman, D. G. 1987. Preservatives Available for Use, In R. G. Board, M. C. Allwood, and J. G. Banks, Eds., Society for Applied Bacteriology Technical Series no 22, *Preservatives in the Food, Pharmaceutical and Environmental Industries.* Blackwell Scientific Publications.
11. Bloomfield, S. 1991. Methods for Assessing Antimicrobial Activity, p. 1-22. In S. P. Denyer and W. B. Hugo, (Eds.), The Society for Applied Bacteriology Technical Series No 27, *Mechanisms of Action of Chemical Biocides.* Blackwell Scientific Publications, Oxford.
12. Gilbert, P., M. R. W. Brown, and J. W. Costerton. 1987. Inocula for antimicrobial sensitivity testing: A critical review. *J. Antimicrob. Chemother.*, 20, 147.
12a. Gilbert, P., P. J. Collier, and M. R. W. Brown. 1990. Influence of growth rate on susceptibility to antimicrobial agents: Biofilms, cell cycle, dormancy, and stringent response. *Antimicrob. Ag. Chemother.*, 34, 1865-1868.
13. CTFA Preservation Subcommittee, 1981. A Guideline for the Preservation Testing of Aqueous Liquid and Semi-Liquid Eye Cosmetics. *CTFA Technical Guidelines*, Cosmetic, Toiletry, and Fragrance Association, Washington, D.C.
14. American Public Health Association. 1980. *Standard Methods for the Examination of Water and Wastewater*, 1st ed. American Public Health Association, Inc., Washington, D.C.
15. U.S. Pharmacopeia Convention. 1990. Microbiological Tests - Antimicrobial Preservatives - Effectiveness (51). In *U.S. Pharmacopeia XXII.* National Formulary. U.S. Pharmacopeia Convention Inc., Rockville, Md.
16. DIFCO Manual page 349.
17. *Ann. Dermatol. Syphilol.* 1892-1893 (DIFCO p771).
18. American Public Health Association. 1978. *Standard Methods for the Examination of Dairy Products*, 14th Ed. American Public Health Assoc., Inc., Washington, D.C.
19. American Public Health Association. 1976. *Compendium of Methods for the Microbiological Examination of Foods.* American Public Health Association, Inc., Washington, D.C.
20. *Am. J. Publ. Health.* 1951. 41, 292.
21. *Bull. d. Inst. Sierotopl, melan.* 1926. 5, 173.
22. *Am. Rev. Resp. Dis.* 1967. 95, 1041.
23. CTFA. Preservation Subcommittee, 1981. A Guideline for the Determination for Adequacy of Preservation of Cosmetics and Toiletry Formulations, *CTFA Technical Guidelines*, Cosmetic, Toiletry, and Fragrance Association, Washington, D.C.
24. Brannan, D. K., 1996. The role of packaging in product preservation. Chap 10, in *Preservation-Free and Self-Preserving Cosmetics and Drugs.* J. J. Kabara and D. S. Orth, Eds., Marcel Dekker, New York.
25. Weber and Black. October, 1948. *Soap and Sanitary Chemicals.*

26. U.S. Pharmacopeia Convention. 1995. Microbiological Tests - Microbial Limit Tests (61). In *U.S. Pharmacopeia XXIII.* National Formulary. U.S. Pharmacopeia Convention Inc., Rockville, MD.
27. Engley, F. B. and B. P. Dey. 1970. A Universal Neutralizing Medium for Antimicrobial Chemicals. p. 100-106. In *Proceedings of the 56th Mid-Year Meeting of the Chemical Specialties Manufacturers Association,* New York.
28. Brewer, J. H. 1940. A Clear Liquid Medium for the "Aerobic" Cultivation of Anaerobes. *J. Am. Med. Assoc.* 155, 598.
29. Quisro, R. A., I. W. Gibby, and M. J. Foter. 1946. A Neutralizing Medium for Evaluating the Germicidal Potency of the Quaternary Ammonium Salts. *Am. J. Pharm.* 118, 320.
30. Sutton, S. W. V., T. Wrzosek, and D. W. Proud. 1991. Neutralization Efficacy of Dey-Engley Medium in Testing of Contact Lens Disinfecting Solutions. *J. Appl. Bacteriology.* 70, 351.
31. *A. J. Clin. Path.* 1954, 24, 621.
32. *Rev. Latinoam Microbiol.* 1958, 1, 125.
33. Hugo, W. B. and S. P. Denyer. 1987. The Concentration Exponent of Disinfectants and Preservatives (Biocides)., p. 281. In R. G. Board, M. C. Allwood, and J. G. Banks (Eds.), *Preservatives in the Food, Pharmaceutical and Environmental Industries.* Blackwell Scientific Publications, Boston.
34. CTFA Preservation Subcommittee., Spring 1973. *CTFA Cos. J.*, 5(1), 2-7.
35. Owen, Emily M. Summer 1969. *TGA Cos. J.* 12-15.
36. American Society for Testing and Materials. 1991. Standard method for water-containing cosmetics (ANSI/ASTM E640-78), *Annual Book of ASTM Standards,* volume 11.04, American Socity for Testing and Materials, Philadelphia, p. 290.
37. Gerhardt, Philipp, R. G. E. Murray, Costilow, N. Ralph, Nester, W. Eugene, Willis A. Wood, Noel R. Krieg, and G. Briggs Phillips. 1981. *Manual for Methods for General Bacteriology.* pages 67, 174, 504. American Society for Microbiology, Washington, D.C.
38. Orth, D. S., C. M. Lutes, and D. K. Smith. 1989. Effect of Culture Conditions and Method for Inoculum Preparation on the Kinetics of Bacterial Death During Preservative Efficacy Testing. *J. Soc. Cosmet. Chem.* 40, 193.
38a. Orth, D. S. 1979. Linear Regression Method for Rapid Determination of Cosmetic Preservative Efficacy. *J. Soc. Cosmet. Chem.* 30, 321.
39. Cosmetic Toiletry and Fragrance Association Preservation Sub-Committee. 1973. Evaluation of Methods for Determining Preservative Efficacy. *Cosmet. Toilet. Frag. Assoc. J.* 5, 1.
40. Zar, J. H. 1984. Simple Linear Regression. p. 261-291. In *Bio-statistical Analysis,* Prentice-Hall, Inc., Englewood Cliffs, NJ.
41. Sutton, S. V. W., R. J. Franco, M. F. Mowrey-McKee, S. C. Busschaert, J. Hamberger, and D. W. Proud. 1991. D-value determinations are an inappropriate measure of disinfecting activity of common contact lens disinfecting solutions. *Appl. Environ. Microbiol.* 57, 2021-2026.
42. Barnes, Mary and G. W. Denton. October, 1969. Capacity Tests for the Evaluation of Preservatives in Formulations. p. 729-733. *Soap Perf. Cosmet.*
43. Brannan, D. K., J. C. Dille, and D. J. Kaufman. 1987. Correlation of *In Vitro* Challenge Testing with Consumer Use Testing for Cosmetic Products. *Appl. Environ. Microbiol.* 53, 1827.

44. John R. Sabourin. December 1990. Evaluation of Preservatives for Cosmetic Products. page 24. *Drug and Cosmetic Industry*.
45. Brannan, D. K. and J. C. Dille. 1990. Type of Closure Prevents Microbial Contamination of Cosmetics during Consumer Use. *Appl. Environ. Microbiol.* 56, 1476.
46. A. D. Russell. 1981. Neutralization Procedure in the Evaluation of Bactericidal Activity, p. 45-49. In C. H. Collins, M. C. Allwood, S. F. Bloomfield, and A. Fox (Eds.). The Society for Applied Bacteriology Technical Series No. 16, *Disinfectants: Their Use and Evaluation of Effectiveness.*, Academic Press, Inc. (London), Ltd., London.
47. Dey, B. P. and E. B. Engley, Jr. 1983. Methodology For Recovery of Chemically Treated *Staphylococcus aureus* with Neutralizing Medium. *Appl. Environ. Microbiol.* 45, 1533.
48. Proud, D. W. and S. V. W. Sutton. 1992. Development of a Universal Diluting Fluid for Membrane Sterility Testing. *Appl. Environ. Microbiol.* 58(3), 1035.
49. Anon. 1991. *ASTM Evaluating Inactivators of Antimicrobial Agents in Disinfectant, Sanitizer, and Antiseptic Products* E1054-91.
50. Jennison, M. W. 1937. Relations between plate counts and direct microscopic counts of *Escherichia coli* during the logarithmic growth period. *J. Bacteriol.* 33, 462-467.
51. Jennison, M. W. and G. P. Wadsworth. 1940. Evaluation of the errors involved in estimating bacterial numbers by the plating method. *J. Bacteriol.* 39, 389.
52. Brannan, D. K. 1995. Cosmetic preservation. *J. Soc. Cosm. Chem.* 46, 199-220.
53. Orth, 1993. Microbiological test methods, in *Handbook of Cosmetic Microbiology*. Marcel Dekker, New York.
54. Ray, B. 1989. *Injured Index and Pathogenic Bacteria*, CRC Press, Boca Raton, FL.
55. Madden, J. M. 1984. Microbiological methods for cosmetics. pp. 573-603, in *Cosmetic and Drug Preservation: Principles and Practices*, J. J. Kabara (Ed.), Marcel Dekker, New York.
56. Balows, A., W. J. Hausler, Jr., K. L. Hermann, H. D. Isenberg, and H. J. Shadomy. 1991. *Manual of Clinical Microbiology*, 5th ed., ASM, Washington, D.C.
57. D. S. Orth. 1981. Principles of Preservative Efficacy Testing. *Cosmet. Toilet.* 96, 43.
58. F. G. Slaughter. 1950. *Immortal Magyar Semmelweiss, Conqueror of Childbed Fever.* p.3. Schuman, New York.
59. M. L. Rotter. 1984. Hygienic Hand Disinfection. *Infection Control.* 5(1), 18.
60. A. C. Steere and G. F. Mallison. 1975. Handwashing Practices for the Prevention of Nosocomial Infections. *Ann. Int. Med.* 83(5), 683.
61. Ayliffe, G. A. J. et al. 1975. *J. Hyg.* 75, 259.
62. L. J. Taylor. January 12, 1978. *Nursing Times*, page 54.
63. P. B. Price. 1938. New Studies in Surgical Bacteriology and Surgical Technique. *J. Am. Med. Assoc.* 111, 1993.
64. A. M. Klingman. 1965. The Bacteriology of Normal Skin. p 16. In H. I. Maibach and G. Hildick-Smith (Eds.), *Skin Bacteria and Their Role in Infection.* McGraw-Hill, New York.
65. M. J. Marples. 1969. The Normal Flora of the Human Skin. *Br. J. Dermatol.* 81(suppl), 15-17.

66. Ayliffe, G. A. J. 1980. The Effect of Antibacterial Agents on the Flora of the Skin. *J. Hosp. Infect.* 1, 111.
67. Ayliffe, G. A. J. 1984. Surgical Scrub and Disinfection. *Infect. Control* 5(1), 23.
68. Noble & Somerville. 1974,(cited in Ayliffe 1980).
69. Gibbs, B. M. and L. W. Stuttard. 1967. Evaluation of Skin Germicides. *J. Appl. Bacteriol.* 30, 66.
70. Selden, R., S. Lee, W. L. L. Wang, et al. 1971. Nosocomial *Klebsiella* Infections: Intestinal Colonization as a reservoir. *Ann. Intern. Med.* 74, 657.
71. Salzman, T. C., J. J. Clark, and L. Klemm. 1967. Hand Contamination of Personnel as a Mechanism of Cross-Infection in Nosocomial Infections with Antibiotic-Resistant *E. coli* and *Klebsiella Aerobacter.* In *Antimicrob. Agents Chemother.* p. 97.
72. Adler, J. L., J. P. Burke, D. F. Martin, et al. 1971. *Proteus* Infections in a General Hospital. II. Some Clinical and Epidemiological Characteristics. With an Analysis of 71 Cases of Proteus Bacteremia. *Ann. Intern. Med.* 75, 531.
73. Lowbury, E. J. L., B. T. Thom, H. A. Lilly et al. 1970. Sources of Infection with *Pseudomonas aeruginosa* in Patients with Tracheostomy. *J. Med. Microbiol.* 3, 39.
74. Rammelkamp, C. H., Jr., E. A. Mortimer, Jr., and E. Wolinsky. 1964. Transmission of *Streptococcal* and *Staphylococcal* Infections. *Ann. Intern. Med.* 60, 753.
75. Mortimer, E. A., Jr., P. J. Lipsitz, E. Wolinsky et al. 1962. Transmission of Staphylococci Between Newborns: Importance of the Hands of Personnel. *Am. J. Dis. Child.* 104, 289.
76. Mortimer, E. A., Jr., E. Wolinsky, A. J. Gonzaga et al. Role of Airborne Transmission in *Staphylococcal* Infections. *Br. Med. J.* 1, 319.
77. Frappier-Davignon, L., A. Frappier, and J. St. Pierre. 1959. *Staphylococcal* Infection in Hospital Nurseries. Influence of Three Different Nursing Techniques. *Can. Med. Assoc. J.* 81, 531.
78. Fleck, A. C., Jr. and J. O. Klein. 1959. The Epidemiology and Investigation of Hospital-Acquired Staphylococcal Disease in Newborn Infants. *Pediatrics.* 24, 1102.
79. Wolinsky, E., P. J. Lipsitz, E. A. Mortimer, Jr. et al. 1960. Acquisition of *Staphylococci* by Newborns. Diecst versus Indirect Transmission. *Lancet* 2, 620.
80. Eisenbach, K. D., R. M. Reber, D. V. Eitzman et al. 1972. Nosocomial Infections Due to Kanamycin-Resistant (R)-Factor Carrying Enteric Organisms in an Intensive Care Nursery. *Pediatrics.* 50, 395.
81. Adler, J. L., J. A. Schulman, P. M. Terry, et al. 1970. Nosocomial Colonization with Kanamycin-Resistant *Klebsiella pneumoniae,* types 2 and 11, in a Premature Nursery. *J. Pediatr.* 77, 376.
82. Burke, J. P., D. Ingall, J. O. Klein et al. 1971. *Proteus mirabilis* Infections in a Hospital Nursery Traced to a Human Carrier. *N. Engl. J. Med.* 282, 115.
83. Newsom, S. W. B., C. Rowland, and F. C. Wells. 1988. What is in a surgeon's glove?, *J. Hosp. Infect.*, 11(supplement A):244.
84. Anon. 1987. Evaluation of Health Care Personnel Handwash Formulation. In *ASTM Standards on Materials and Environmental Microbiology.* E1174-87.
85. Ayliffe, G. A. J. 1989. Standardization of Disinfectant Testing. *J. Hosp. Infect.*, 13, 211.

5 VALIDATION OF METHODS

DANIEL K. BRANNAN

CONTENTS

INTRODUCTION

Within this chapter, we offer general guidelines for validating and documenting microbial methods for the personal care product industry. Both the plant and laboratory have areas needing validation. By instituting a rigorous program of validation, we can ensure that the methods and systems we use do what they purport to do and thus we can have a high degree of confidence in them. Validation should be an integral part of continuous quality improvement in order to provide our customers a high-quality finished product.

Validation of microbiological procedures within the personal care product industry simply establishes their efficacy, accuracy, and reproducibility. Validation is done whenever a new procedure is developed and whenever significant changes are made to existing systems, methods, and procedures. Validation without documentation, however, is useless; the results of the validation should be documented in an organized and permanent record.

0-8493-3713-5/97/$0.00+$.50

A MODEL FOR VALIDATION

There are five basic steps in any validation program. In order to enable better understanding of these steps, the validation of the preservative efficacy test (PET) is used as an example.

1. Define what the system, method, or procedure is supposed to do (e.g., a PET should predict the rate of consumer contamination of a product).
2. Identify and control, if possible, the variables in the system, method, or procedure (e.g., abuse of the product during use cannot be controlled).
3. Establish acceptance criteria for the system, method or procedure *before* beginning the validation process (e.g., well preserved products should not become contaminated under ordinary use and foreseeable abuse).
4. Develop and execute a protocol to determine if the procedure meets the criteria.
5. Document the procedures and results.[1]

Documentation includes information on how the procedure works, as well as how the data is obtained. For example, when validating a sanitization procedure, one would document the date of sanitization, the equipment being sanitized, the product for which the equipment was being sanitized, the sanitizer and its concentration or method of sanitization used, the performance criteria for acceptance, whether or not specifications were met, and appropriate signatures of witnesses, and proof that the results were properly reviewed by qualified supervisory personnel. Records are generally kept from 3 to 7 years, although this criteria may be variable.

There are three approaches to validation: prospective, concurrent, and retrospective. To provide an example of these approaches, we can again use the preservative efficacy test. A prospective approach involves qualification of the system and subsystems using an approved protocol with appropriate acceptance criteria that is executed and the data analyzed and reported for approval or rejection. There is a published example of this approach.[1] A concurrent approach to validation is based on information gathered during actual implementation of the process or system. This is the type of data generated by periodically allowing consumer panels to use the product that has passed the preservative efficacy test and determining if the product is still uncontaminated. The last approach is retrospective validation. In this type of validation, one reviews and analyzes historic data. For example, one could determine if the PET was valid by whether or not consumer comments/complaints were related to microbial growth in the product. This assumes, of course, that the product samples with comments potentially related to microbial contamination were assayed for microbial content.

VALIDATION OF EQUIPMENT CLEANING AND SANITIZATION

Validation, as applied to cleaning and sanitization, is the evaluation of data gathered under a validation scenario (prospective, concurrent, or retrospective) that assures the procedures produce acceptably cleaned and sanitized equipment surfaces that contact the product.

Most prospective validation schemes of sanitization validation require that a test liquid be used to mimic the product while doing the validation run. After cleaning and sanitization, the sterile unpreserved liquid is introduced into the equipment and put through the same conditions one would expect for the product being made on the system. After contacting the same surfaces to which the product would be exposed, an aliquot is taken and examined for microbial content. Decisions are made regarding the adequacy of the sanitization procedure based on the content data with 10 to 100 CFU/ml being common limits. Some organizations even purposely contaminate the system with a known amount of organisms, followed by a sanitization and then running the test liquid (sometimes microbiological media) through the system.

Aside from the obvious risk of introducing microorganisms unnecessarily into the system, another danger exists when the system passes a prospective validation. It gives the personnel a false sense of security. Validation is a snap-shot of a system that is being subjected to a continuous process of biological adaptation and evolution. Thus, when a validated system gets contaminated, the people are incredulous that a problem occurred on a "validated system." The missing piece of information is the variable that can never be controlled: biological evolution which allows for microbial adaptation to the biocides being used. The other uncontrolled variable is the human element, one which is characteristically uncontrolled.

In conducting a concurrent validation of sanitization, one generally relies on swab samples of critical control points in the system before and after a sanitization during actual production. One can even analyze final washout water to remove chemical sanitizers as in prospective validations if the water is free of microorganisms. The uncontrolled variables here are whether or not all the critical control points have really been sampled. This approach assumes that no dead-legs are in system.

In conducting a retrospective validation of sanitization, one relies on the historical production data of clean product and correlates that production data with sanitization documentation. The uncontrolled variable here is that this approach does not account for the biocidal activity of the product masking any sanitization deficiencies.

The details of sanitization validation include the cleaning method used before sanitization, the sanitization method, the type of sanitizer employed (if chemical), its use concentration and temperature, contact time of the sanitizer, type of equipment or surface sanitized, and the acceptance criteria. Generally, the best sanitizer to use is heat. Use of at least 180°F (80 to 85°C) for up to one half hour is highly effective, assuming the heat penetrates through the pipes entirely. Heat destroys biofilms quite effectively. After heat, chlorine at 200 ppm free chlorine is next effective, assuming the system has been thoroughly cleaned out.

The validation program and even the selection of sanitization procedures do not have to be done by the microbiologist alone. Instead, he or she should facilitate their implementation by including a variety of people such as the safety officer, engineers, production people, quality assurance, and upper management. The microbiologist should drive the process but should have these other people involved in order to have complete acceptance and compliance with the standards. Revalidation should also be considered anytime a substantive change takes place either in the sanitization procedure, equipment, or process.

Documentation is the most difficult step simply because of the stigma associated with paper work. However, without documentation, the procedures effectively were never done. A properly documented sanitization validation includes the date of sanitization, the equipment sanitized, the name and batch number of the product in the equipment, the type of sanitizer used and/or the sanitization procedure used, its use concentration, signature from person performing the procedure and date, and the counter signature of the supervisor of the shift in which the equipment was used.

VALIDATING THE CLEANLINESS OF THE PLANT ENVIRONMENT

Two areas are usually analyzed for determining the microbial quality of the plant environment: air and surfaces. Air sampling equipment is fairly complex, and the need for it in a cosmetic plant is rarely warranted. Where used, the ability of the sampler to recover organisms of interest without hindering their growth should be validated. More often used are settling plates. The big issue here is the effect of drying on the ability of the plate to support growth. Known amounts of organisms in an aerosol could be sprayed on the plates to validate them. Surface monitoring can be validated by attempting to recover known numbers and types of microorganisms from the various types of surfaces. Various surface monitoring techniques such as swabs or Rodac plates may be used.

Documentation in the environmental control program involves keeping track of trends in contamination levels in the environment using process control charts. Significant spikes or deviations may indicate the need for action. The data needed to provide for proper documentation are date, time, and location of the sample, method of sampling, who did the sampling, lab procedures used including the growth medium and the time/temperature of incubation, the data, and any additional pertinent comments such as general activity at the time of sampling, presence of visible dirt or airborne dust, presence of air currents from air conditioning vents, and date of last surface cleaning samples.

VALIDATION IN THE MICROBIOLOGY LAB

Laboratory-generated data are used in quality control and assurance. The data must be reproducible and reliable. This is accomplished by controlling laboratory operations with the provision of well-documented and validated procedures. These procedures permit safe, efficacious, and reliable production of the consumer product.

Media

Microbiological culture media must contain available sources of carbon, nitrogen, and appropriate trace elements. Fastidious organisms may also require vitamins or other growth-factors. The growth of microorganisms in a medium depends on proper preparation. Documentation should include dates of receipt and preparation, pH, lot number, and any deviations from performance standards outlined by the manufacturer in order to provide reliable test results.

As each batch of media is made, a record should be kept including such information as the manufacturer's batch number, the date and amount made, the final pH and any adjustments if needed, how it was sterilized and the results, the ingredient lot numbers for media requiring separate ingredients, when a medium expires, the test organisms to show the medium promotes growth and the results, and the signature of the preparer. Perhaps the one raw material most rarely checked or validated is the water used in media making. Purified water obtained by distillation or ion-exchange treatment should be used. The quality of this water should be checked chemically, at regular intervals, to determine its suitability.

Growth support checks of a medium should consider which organisms you are wishing to detect. Selective and enrichment media need to support and detect their appropriate organisms. Media for identification of microorganisms should include positive and negative control organisms compared to uninoculated controls. Validation also includes establishing storage conditions and shelf life limits. These limits should be determined by appropriate performance testing.

As examples of the above principles, we can follow a typical lab's systems for validation of media. The purpose of a medium is to detect organisms in a sample. Some of the variables to control are sterilizing time, preparation methods, storage time and temperature, and incubation time and temperature. The key criterion for acceptance is that the medium support the growth of the organisms we wish to detect without inhibition (even in the presence of the product) of growth; the medium should also be sterile. To conduct the validation protocol for sterility, one can rely on the validation of the autoclave (see below) and couple to this, the demonstration that uninoculated media do not show growth.

In validating the microbial content test, for example, one can combine the validation of neutralization and growth support in one test. In the sample data provided below, distilled water is used as a control for the particular product being validated against. The procedure calls for the organisms to be added to 90 ml of Letheen Broth neutralizer such that they will be suspended at a concentration of about 100 CFU/ml. Then the product or the water control (10 grams) is added to the broth and mixed immediately. This order of addition is important. We must ensure that the neutralizer eliminates the biocide activity of the product before the biocide activity can eliminate the organisms. After the mixing, 0.1 ml aliquots may be spread plated onto the agar surface or 0.5 ml aliquots may be put into melted agar deeps, vortexed and poured. Alternatively, 0.5 to 1.0 ml aliquots may be placed into plates and melted agar poured on top of the aliquot and swirled to ensure complete mixing. Whichever method of plating is chosen, it should validated against the same method using the water control and compared to the other methods to show equivalency if various labs use different plating protocols. Data such as those seen in Table 5-1 of the validation register below are typical of spread plating. The data are expressed on a CFU/g product (or water) basis.

The data can be subjected to statistical testing using a standard t-test, provided the count values (CFU/g product) are first transformed logarithmically (log x+1) and the data are worked using the log transformations.

One of the classic ways of validating growth support and neutralization (ASTM method) is streaking a plate that shows no growth after product has been added to it with test organisms. If the streak grows, then the medium is considered validated with respect to growth support. This approach is not valid. Growth only shows that the biocide was finally inactivated at the time the streak was done, not at the time the product was plated. No growth on the plate does, however, prove that the product was never inactivated to allow microbial growth.

Microbial Content Tests

Microbial content tests are needed to determine the bioburden of raw materials, package components, water, and the final product itself. The microbial content tests typically use plate count procedures that should be evaluated for their reproducibility. In particular, the procedures, materials, and equipment used need to be shown effective. The procedures should be written in a standard operations manual in sufficient detail that someone educated in the science of microbiology could follow.

TABLE 5-1

MICROBIOLOGY VALIDATION REGISTER

DATE:	12/96
PRODUCT:	XYZ Brand Shampoo Batch no. 96AA
MEDIA:	Letheen broth and TSA-Tween
OBJECTIVE:	To ensure that the neutralizer and plating media used recover organisms regardless if they are in the presence of biocides or not.
CONDITIONS and PROCEDURES:	Media autoclaved at 121°C for 20 minutes. Incubations done at 30°C for 48 hours. Add organisms to 90 ml neutralizer at 100 cfu/ml concentration. Add 10 grams of product (or water as a control) to the 90 ml of inoculated neutralizer. Mix thoroughly and remove ten 0.1 ml aliquots using an Eeppendorf pipet into plates to be spread plated.

DATA/RESULTS:

Escherichia coli			*Pseudomonas aeruginosa*			*Staphylococcus aureus*		
Product		**Control**	**Product**		**Control**	**Product**		**Control**
138		149	100		119	60		28
188		149	120		115	51		41
102		134	119		109	40		47
99		166	115		118	93		46
131		113	132		163	21		47
117		137	124		134	28		76
146		137	108		133	53		17
115		147	122		131	49		58
138		120	97		108	56		53
152		124	125		77	31		50
133	Average	138	116	Average	121	48	Average	46
$p = 0.62$; $t = 0.512$			$p = 0.58$; $t = 0.570$			$p = 0.814$; $t = 0.232$		

CONCLUSIONS: Counts recovered from the test diluent and medium in the presence of the product did not differ significantly from those obtained in the presence of water. Thus the methods are valid for use.

Note: Plate Counts to Validate Recovery of Organisms by Spread Plate Using TSA-Tween and Letheen Broth Neutralizer. Data are in CFU/g.

Some microbiological procedures are compendial in nature (e.g., USP), others are guidelines (e.g., CTFA), and some are collaborative lab methods (e.g., AOAC). Regardless of their apparent "official" nature, they are still subject to considerable human error and opinion on the part of the panel developing the method. As a result, they should not be trusted as validated methods, particularly since none of the published methods are designed specifically for the product that a particular manufacturer makes.

Thus, the method used for determining microbial content must be validated by the manufacturer for each of its products or class of products. When validating a microbial content test, the elements to be validated include plating methods (e.g., spread or pour), dispersion techniques (mechanical or manual), incubation time and temperature, dilution factors, and neutralizing agents. An example of how this can be done was provided in the discussion above on media and growth support validation.

Identification

Organisms from the plate count are identified based on colony morphology, Gram reaction, pigment production, motility, and biochemical characteristics including nutritional requirements and reaction on selective and differential media. Validation of the methods used to identify these organisms include the ability to do the Gram stain properly, and to conduct the biochemical assays properly. The use of standard slides with known Gram positive and negative organisms on them can help validate a technician's ability to do the Gram stain. Use of reference cultures on the media used to identify the organisms is appropriate to validate the biochemical assays. In many cases, microbiology labs rely on the more modern rapid identification methods. Despite their extreme reliability, they must still be validated using known reference cultures to prove they are capable of identifying the organisms of interest.

Laboratory Equipment

Laboratory equipment should be maintained so that it will perform properly. Any performance deficiency and corrective action taken should be documented. Some of the items needing validation and documentation include buffer solutions for pH meters, thermometers, temperature gauges, and sterility indicators such as spore strips for autoclaves. A file for each piece of equipment to be validated should be kept and should contain the following information: the model and serial numbers, purchase date, manufacturer, maintenance and operation manuals, and technical service representatives.

The operator's manual on each piece of equipment should establish control steps for carrying out validation. Documentation should include the frequency and criteria for acceptance, performance standards, any deficiencies, and the corrective action taken. Specialized checks, such as calibration of centrifuges, balances, and scintillation counters, should be performed by an authorized representative of the manufacturer.

Installation of the piece of equipment should be validated in order to establish basic operational requirements, specifications, and tolerances. Most equipment manufacturers provide recommended installation procedures, but do not give qualification instructions to validate that the equipment was installed properly. The user is left to determine what validates proper installation. In validating installation, one should determine what are the critical features of operation that might affect function, variability, data, and records. One should also consider the effect that usage patterns have on the equipment; usage may be continuous, intermittent, variable, or with long down times between uses. Idled equipment usually needs more attention than that in constant operation.

Calibration of the installed equipment should take particular care when selecting the reagents used with the machine, the techniques for detection of variances, and the measurement of what constitutes control. Wherever possible, the calibration program should use recognized standards, especially those from recognized standard-setting agencies or associations.

Once the equipment is validated as operational, it will require periodic monitoring to detect variation in set standards of performance. Major variations of these standards may require equipment shut-down and revalidation. Use of control charts is a good statistical control process to verify that the equipment continues to operate within limits. A written standardized procedure is required for all laboratory equipment. Of course, all validation procedures require attestment of at least two signatures (with dates) to the records kept in a bound laboratory notebook or forms. These validation records should be placed on the equipment or instrument or may be in a centralized location.

Also included in validation of equipment is the need for preventative maintenance to provide evidence that the equipment is in a state of control. Preventative maintenance helps minimize malfunction and high variation. It provides stable laboratory operations and is preferable to sporadic, uncontrolled major maintenance. Routine maintenance involves daily, weekly, or monthly activities such as lubrication and cleaning and replacement of recording paper. A written record is required showing the date and time when the maintenance was performed. When more skill and technical knowledge is required for maintenance activities than the average technician can handle, specially trained individuals perform this function.

The key elements to control and validate for refrigerators and freezers is that these maintain the temperatures for which they are used. The sophistication of their electronic controls will decide the validation steps needed. Common temperature controlling devices that are simply rheostats only set the instrument at a certain point without feedback. Refrigerators and freezers with these controls are dependent somewhat on the ambient temperature of the environment into which they are placed. Validation must take into account the potential for variation of the ambient temperature, the amount of opening of the doors of the instrument, and the temperature gradients that may exist within the refrigerator or freezer itself whether empty or full. Validation procedures should check the internal chamber temperature with a calibrated thermometer traceable to an NBS standard or with calibrated temperature sensors independent of the equipment's operational devices. Alarms are good to warn of equipment malfunction. When very precise temperatures are needed, more highly specialized control devices and feedback systems can be used.

Centrifuges may be used to harvest inoculum for a preservative efficacy test or enough cells to set up seed-lot cultures for maintaining the organisms. One of the key areas for control is the velocity of spin. This should be kept even and consistent at each of the speeds one wishes to use. Therefore, tachometers need to be kept calibrated in order to check the dial settings of the centrifuge with the actual spin velocity. In refrigerated centrifuges, the temperature variability within the chamber must be determined. In addition, procedures for which materials (e.g., materials and solvents) are allowed and which are prohibited in the centrifuges should be maintained, along with procedures to ensure they do not remain in the centrifuge head after its use. Use of timing devices must also be validated for reproducibility and reliability on the centrifuge (see below).

The elements to control in blending and mixing devices include velocity and timing. Therefore, measuring blending and mixing capabilities should be done at several velocities. Timers should be checked against calibrated timers.

Maximum/minimum thermometers are used to control and monitor temperatures in equipment. Using these kinds of thermometers, the operator will manually check the thermometer and determine the maximum temperature reached in the instrument over a period of time. Max/min thermometers may be used in rooms, chambers, freezers, refrigerators, incubators, and water baths. The criticism of these thermometers is that they do not show how long the maximum temperature was held. Calibration of any thermometer can be done by comparing the thermometer with an NBS calibrated thermometer at two control temperatures that span the range desired.

Balances used in microbiology labs range from simple triple beam balances to electronic balances that can tare and calibrate at the push of a button. Regardless of type, all balances must be properly calibrated with certified weights. Some balances are calibrated each day. The modern electronic balances require calibration less frequently, but usually at least monthly with a single weight.

The pH meter also comes in a diversity of designs. The manufacturer's recommendation for calibration should be followed. At a minimum, daily calibration against two buffer solutions should be done, one at pH 7 and the other (the slope) at either pH 4 or 10 depending on whether the solutions to be measured are acidic or basic.

Timers are used in all kinds of laboratory processes and for all kinds of equipment and instruments. Calibrating these timers is often overlooked. However, certified calibrating stop watches exist to check timers.

Instruments that measure turbidity include spectrophotometers, Klett meters, and nepholometers. These instruments are complex enough to demand that the operations manuals are carefully followed for installing, operating, and calibrating. Typically for a spectrophotometer, a blank is used to provide the standard for 100% transmittance (or a 2.0 optical density) and complete blocking of the light at a specific wavelength is provided as the standard for 0% transmittance (or a 0 optical density).

Incubators provide controlled temperatures and moisture for the promotion of bacterial growth in inoculated media. These have specific details for their operation and precision. Installation, operation, and calibration should consider the variability acceptable for reproducible reliable data collection. For example, the temperature variation within the incubator should be documented since extremes can exist from top to bottom of an incubator and can be different depending on how full or empty the incubator is. Many incubators resolve this problem by the use of fan-circulation of the air inside the incubator. Use of multiple thermocouples throughout the incubator and hooked into a computer is an efficient way of determining the temperature profile of the incubator. If this is beyond the ability of the lab, the manufacturer of the incubator should have such capacity. In sophisticated incubators, temperature data can be recorded continuously on charts or in computer data systems.

In a sense, water baths are similar to incubators. How much variation in temperature is permissible depends upon for what use the water bath is put. In many ways, the analogy to incubators is accurate since circulation of water (rather than air) will provide better temperature control in water baths and calibration of the thermal sensors and fluid flow control devices are the basic elements of control. In addition, thermal mapping should also be done during the installation. Alarm systems are available and continuous recording systems can be integrated into the water bath.

Colony counters require minimal controls. Nevertheless, written procedures are needed for their proper use and maintenance. The most critical element of the colony counter is the mechanism for recording the colony counts. Manual tabulators are still very common but are being replaced by the touch probe to eliminate human variability. In one case, a "clicker" is used where the plunger is pressed with the thumb each time a colony is counted. In the touch probe, a count is recorded each time the probe is pressed to the colony. Variability is controlled by providing a standard plate with a known number of colonies and having the technician count them. The more sophisticated automated surface colony readers take away the human elements of fatigue in counting and errors due to lack of visual acuity. They are used where high numbers of plate counts are done routinely. The automated colony counters are based on either image analysis or laser interruption as it passes through a colony. These counters must be standardized using plates with patterns of spots to mimic colonies painted on them. Installation qualification and employee training is critical for operating scanning recording colony counters.

Laminar flow hoods are used for culture transfers and for putting up cultures permanently to ensure culture purity. Standards for laminar air flow hoods focus on maintaining laminarity and, of course, sterility. Both of these measures can be checked daily. In addition, detection of leaks in the HEPA filters is critical to operation of laminar flow hoods. A simple mechanism to validate the sterility of the hood is to use either settling plates or air samples from within the hood. In addition to having written procedures for this validation and the operation of the hood, detailed instructions for the personnel operating the laminar flow hood should be provided so the operators themselves do not compromise the flow of air.

Sterilizers

Laboratories require the use of sterilizers to provide sterile media and materials for assaying the microbial content of products. The most commonly used sterilizer is the autoclave. It operates by killing the microbes due to exposure to moist heat through the use of steam under pressure. The critical function is to ensure that this moist heat penetrates throughout the load to destroy all organisms. One of the key variables to consider is the size and type of load. Thus, validation of the autoclave should include instrument or component calibration, installation qualification, operational qualification, and certification under a diversity of loads.

During installation and operational qualification, a thermal profile of the empty chamber and performance of various load profiles with thermal sensors and biological indicators provides the validation needed. As these operations are performed, one should identify the chamber area that takes the longest to reach temperature before sterilization takes place.

The simplest way to do this is to use thermocouples that are passed into the autoclave through the chamber door. These are hooked into recording devices that are microprocessors with computer tie-ins. After this initial thermal mapping, one should mimic typical loads expected in the planned sterilizer operations. The type of load will impact the thermal profile. Absorptive loads may cause steam to condense and thus take longer for the chamber to become saturated with steam. Fluid and solid loads may take longer to heat up. The type of container that the material is in will also affect the efficacy of the autoclave. For example, a plastic flask will not permit penetration of heat to the media as quickly as a glass flask. The most common standard in microbiological laboratory operations includes the one for making microbiological media in the autoclave: "sterilize for 15 minutes at 121°C." Unfortunately, this standard does not assure sterility. The neophyte often misinterprets this as a sterilizing cycle of 15 minutes at 121°C instead of a thermal exposure of the center of the item to be autoclaved for 15 minutes.

This problem is especially acute when sterilizing large volumes of liquid. These have a high capacity and must be heated up to boiling before being autoclaved. In smaller volumes of 1 liter, for example, the microbiologist can often incorporate this step into the autoclave cycle by extending it to 30 minutes. More often, however, the volume is so great that incorporating the heating time into the sterilizing cycle is not possible. Included in this consideration is the fact that once in the autoclave, the material cannot be stirred and so uneven heating and burning of the medium may take place.

In the microbiology lab the most expedient way to validate the sterilizing cycle is the use of the biological indicator (BI).[3] We have not discussed other sterilizers beyond the autoclave. But the BI is useful for all sterilizer cycle validation including steam, dry heat, or ethylene oxide. For more extensive treatment on the various other sterilizing devices, one may refer to Gardner and Peel.[4] Biological indicators come in a variety of forms, including inoculated product with sterilant resistant spores, spore paper strips or disks, or commercially available self-contained BI systems. These BIs should be placed in each corner of a load and in the center of the autoclave. These are then incubated after exposure to the autoclave to see if the sterilizing was effective.

Decontamination

Relatively little attention is given to decontamination procedures for microbiological cultures and lab benches and materials that may become contaminated during testing of products. The goal of decontamination is usually defined as to render the area and materials as safe to handle. When validating the ability of the autoclave to sterilize the spent cultures, the same kinds of guidelines used to validate the autoclave itself can be employed by either using the cultures themselves or at least using a BI in the center of the load. Essentially, the same principles used to attain sterility should be applied to decontamination processes as well. Decontaminating the lab benches can be validated using contact plates or swabs to monitor the efficacy of the process used. In either case, documentation is needed to track the results.

SUMMARY

While this chapter has of necessity been very general, the guidelines given should provide at least a model for developing the lab's own specific procedures for validation. There really is no one right way to do validation. There is, however, a process to follow where one: (1) defines what the system, method, or procedure is supposed to do; (2) identifies and controls the variables in the system, method, or procedure; (3) establishes acceptance criteria for the system, method, or procedure *before* beginning the validation process; (4) develops and executes a protocol to determine if the procedure meets the criteria; and (5) documents the procedures and results. These steps, if followed, will validate any system.

Regardless of the methods used to validate a process, the one common requirement is documentation. Documentation provides the evidence that justifies the methods used. A simple way to do this is to use a validation register. External documentation in the form of published, *peer-reviewed* scientific journal articles should be done whenever possible.

REFERENCES

1. Brannan, D. K., Dille, J. C., and Kaufman, D. J. 1987. Correlation of *in vitro* challenge testing with consumer-use testing for cosmetic products, *Appl. Environ. Microbiol.*, 53, 1827.
2. Halleck, F. E. 1978. Thermal solution sterilization, *Pharm. Technol.*, June.
3. Pflug, I. J. and G. M. Smith. 1977. "The Use of Biological Indicators for Monitoring Wet-Heat Sterilization Processes." In *Sterilization of Medical Products.* (Eds. E. R. L. Gaughran and K. Kereluk), New Brunswick, N.J., Johnson and Johnson, pp. 193-230.
4. Gardner, J. F. and M. M. Peel. 1991. *Introduction to Sterilization, Disinfection, and Infection Control.* 2nd ed. Churchill Livingstone, Melbourne.

SECTION THREE: PRESERVATION OF COSMETICS

6 PRESERVATIVE DEVELOPMENT

Philip A. Geis and Richard T. Hennessy

CONTENTS

INTRODUCTION

Congress defined cosmetics as "articles applied to the human body for the purpose of cleansing, beautifying, promoting attractiveness or altering its appearance."[1] These personal care products are very sophisticated and diverse in formulation. They often include a variety of natural and synthetic ingredients used to satisfy the aesthetic desires of the consumer. Those desires, and the ingredients, are continuously expanding in number and type. Unfortunately, these ingredients also provide the pH, moisture, and nutritional conditions that support microbial growth. Due to this potential susceptibility, many cosmetics include chemical preservatives to prevent spoilage.

The ever-increasing complexity of cosmetic products offers manufacturers the challenge of identifying stable preservative systems that prevent microbial contamination. They must also insure that these preservative systems are effective throughout the product's life. In this chapter, we

0-8493-3713-5/97/$0.00+$.50

define the factors important to cosmetic preservative testing. We also review current and potential challenge methods that may be useful predictors of a cosmetic's preservative.

Why Do We Preserve Cosmetics?

Microbial contamination of cosmetics is a substantial risk to product quality, regulatory compliance, and consumer health. Manufacturers invest considerable effort to maintain product quality by reducing the risk of microbial contamination. Such efforts include showing effective chemical preservation, controlling raw material quality, and following good manufacturing practices.

The results of microbial contamination include undesired changes in product odor, color, viscosity, and performance.[2] These changes are due to microbiological breakdown of product components and the production of microbial metabolites. Such changes can occur swiftly if the invading microbes have the right environment for rapid reproduction.

More often, microbial contamination is not that obvious. There is usually no immediate visible evidence of a growing population established within the product. Many months may go by before the incipient microbial contamination shows a recognizable change in the product.

The most important result of microbial contamination in cosmetic products is the risk to consumer health. Manufacturers' efforts are usually effective in preventing cosmetic contamination. Health-related incidences from cosmetics are very rare. Contaminated cosmetics have historically been associated with hand infections,[3] severe eye infections,[4–6] and with the death of one immunocompromised individual.[7].

Microbiological safety of a cosmetic is a primary concern to the manufacturer. It is also a regulatory responsibility of the U.S. Food and Drug Administration. By the Food, Drug, and Cosmetic Act,[1] Congress charged FDA with the responsibility to protect the public against adulterated cosmetics. In its microbiological application, the law defines adulterated products as those containing injurious or pathogenic microorganisms or their metabolites.[1,8,9] A determination of adulteration may prompt the Agency to request a product recall or even affect a seizure of the contaminated cosmetic product. These actions have significant economic consequences for the manufacturer.

Strategy for Preservation

Proper design and formulation of an adequately preserved product considers the spectrum of variables that might compromise the product's microbiological status. The presence of an effective and stable chemical preservative system is an important and obvious factor, but it should not be a replacement for good manufacturing practices. Raw material quality, package design, shelf life, and consumer use (as well as foreseeable misuse) are also important elements to the microbiological integrity of a product. Proper control and balancing these elements is a good strategy for maintaining the microbiological quality of a product.[10]

Several factors are important in product protection. The factor under the most immediate responsibility of the microbiologist is the development of preserved cosmetic products. These products should be hostile to microbial growth or survival. For cosmetics, preservation is a balanced technology. It must show adequate antimicrobial hostility to prevent microbial contamination. But it must do this without causing adverse toxicological effects.[11,12] This must all be done without high costs.

We do not get adequate preservation by merely formulating the product to include an antimicrobial chemical. Even disinfectant products formulated specifically to kill microorganisms may still be susceptible to contamination.[13,14] Some have even been the subject of at least one recall.[15] It is therefore important to confirm that the preserved product withstands the types and concentrations of microorganisms it might meet during manufacturing and consumer use.[10,16]

Microbiologists must follow a holistic process in formulating a product to provide stable antimicrobial hostility. The microbiologist then confirms the hostility or resistance to microbial contamination. The central element of this process is the preservative challenge test. Although simple in design, it is complex to perform. Such tests typically consist of four primary operations. Preparation of the product sample is the first step. Then, there is the selection and addition of the microorganisms to the product. We then inoculate product and incubate hold the product while the inoculum incubates. Finally, we estimate the surviving microorganisms and compare those surviving numbers to data for acceptable product.

The tests in current use are either long-term (typically 28 days) challenge tests or rapid tests. Some companies use the latter tests to screen formula and preservative combinations to identify potential apparently effective combinations. The long-term tests are confirmatory and involve the addition of challenge microorganisms and incubation over a period that parallels consumer use patterns.

ELEMENTS OF PRESERVATIVE TESTING

An ideal preservative test shows the relative degree of product antimicrobial hostility. Such tests should reflect the risks of product contamination from manufacture and from consumer use and anticipated misuse. Such tests should give the product development team a tool to assess the microbiological integrity of a cosmetic throughout its product life.

Unfortunately, no such comprehensive test yet exists. Those predictive laboratory techniques that are now available are not without their deficiencies. However, the long record of preservation success established by the cosmetic industry shows that the proper use of current methods and the right manufacturing and packaging controls effectively prevent microbial contamination.

In the following sections, we discuss the long-term (28-day) tests of the types proposed by the Cosmetic Toiletry and Fragrance Association,[17,18] the American Society for Testing and Materials,[19] and the U.S. Pharmacopeia.[11] The principles discussed apply to virtually any preservative challenge test. In this discussion, we will consider the preservative test in its four primary functions. These include preparation of the product sample, inoculation, recovery of microorganisms, and comparison of the results with established acceptance standards.

The Production Sample

The product sample tested to confirm antimicrobial activity should replicate formulation, packaging, and other factors significant to preservation.[10] All raw materials should be of the same quality and from the same source when testing preservation options.

Later changes in formulation such as pH require re-testing. Even minor composition changes of individual ingredients can adversely affect the hostility of a product. For example, the type and quality of treated water affects a product's antimicrobial activity. Changing suppliers of raw materials may provide the presence of trace amounts of other chemicals that require reevaluation to prove continued efficacy.

To address potential product changes resulting from scale-up to larger volume manufacturing, the product produced and packaged for retail sale must be tested. Finally, the preservative test should include unpreserved (susceptible to microbial contamination) and adequately-preserved control products. This allows a test of the inoculum to be sure it is sufficiently robust to provide an appropriate test of preservative efficacy.

The microbiologist has a wide selection of preservatives for use. This menu is supported by the large amount of published information describing specific antimicrobial agents, their compatibilities, and their stability with chemicals and conditions encountered in cosmetic products.[20,21] However, these data are not complete. They often do not address the slow and nonspecific loss of antimicrobial activity seen in some preserved formulations during long-term storage.[22,23]

The microbiologist should confirm the long-term stability of all preserved formulations (a process often accelerated by higher temperatures). This should be done by microbiological and, where methods exist, analytical testing of product stored under ambient and accelerated-aging (usually, elevated temperature) conditions. The results from such testing should confirm the stability of the product's preservative system through conditions of anticipated shipping, storage, and shelf life.

The test container or vessel in which one performs the stability test should be consistent with the package (same composition, capacity, dispensing mechanism) used in retail distribution. We base this recommendation on the finding that some product container materials diminish antimicrobial activity and even contribute to contamination.[24–27]

Most published preservative methods test full-strength (100%) product.[2] However, Cooke et al.[28] recommend challenge testing (of paints) of both 100% product and diluted samples (aqueous dilutions up to 50%). The rationale is that testing of the latter samples reproduces manufacturing conditions that expose diluted product to potential microbial adaptation and growth. Microorganisms adapted to dilute product may acclimate to grow in full-strength product.[28] Therefore, this method provides a worst-case estimate of risks associated with inadequate manufacturing controls.

Addition of Microorganisms

Selection and preparation of an appropriate microbiological challenge inoculum is probably the most critical factor in determining the validity of a preservative test. Each of the long-term challenge tests specifies a core set of inoculum microorganisms (Table 6-1). These include Gram-positive and Gram-negative bacteria and fungi (both those with yeast and exclusively mycelial morphologies). Specific strains of these microorganisms may be obtained from the American Type Culture Collection (ATCC).

A recent survey of cosmetic manufacturers[29] showed that test inocula typically include these organisms. However, most manufacturers also reported the use of resistant strains recovered from inadequately preserved product samples and raw materials, or isolated from the manufacturing sites.

Microorganisms isolated from the manufacturing environment are usually more resistant to preservatives and other hostile conditions than ATCC strains. They include Gram-negative bacteria of the genera *Enterobacter, Serratia, Pseudomonas, Klebsiella,* and fungi of the genera *Fusarium, Penicillium,* and *Saccharomyces.*[29] Each of these strains or isolates typically have a resistance or tolerance to specific antimicrobial chemicals (especially a preservative) or physical conditions (e.g., A_w). Such isolates have increased metabolic versatility[30] and proven adaptation potential.[31,32] Thus, they offer a more robust challenge to any preserved formula and are useful additions to any preservative efficacy test.

The ATCC cultures may or may not have enough biocide resistance. The ATCC is a repository for researchers to deposit cultures. No attempts to verify the claims of the depositor regarding the origin of the culture are made. No attempts to maintain the resistance of the culture are made except to preserve the strain deposited. If more culture requests are made than the original set of vials preserved, then a remaining vial may be subcultured. This represents a departure from the originally deposited culture. If the requests are very high, several subcultures may occur. If enough subcultures occur, the risk of losing the original tolerance is high due to loss of tolerance factors.

Challenge inocula may include broth-grown cultures or those harvested from solid media and suspended in saline.[11,17–19] The growth phase

TABLE 6-1

LONG-TERM PRESERVATIVE EFFICACY PROCEDURES FOR LIQUID COSMETICS

Method	Test Organisms (ATCC No.)	Challenge Procedure	Sampling Intervals	Acceptance Criteria
USP	*Staphylococcus aureus* (6538) *Escherichia coli* (8739) *Pseudomonas aeruginosa* (9027) *Candida albicans* (10231) *Aspergillus niger* (16404) Other likely contaminants	Pure culture method No rechallenge indicated	Test at 7, 14, 21, and 28 days after inoculation	Bacteria: 0.1% of initial number by day 14 and further reduction or no increase thereafter Fungi: initial level by day 14 and no increase or continued reduction thereafter
CTFA	*S. aureus* (6538), *S. epidermidis* *E. coli, Proteus* sp. and other enterics *P. aeruginosa* (15442 or 13388), other Pseudomonas spp. *C. albicans, C. parapsilosis* *A. niger* (9642), *A. fumigatus, Penicillium luteum* (9644), *Fusarium solani* *Bacillus* spp. (optional) House resistant strains	Mixed culture method with use of some subgroups Rechallenge may be performed but not necessary	Test at 0, 1, 7, 14, 21, and 28 days after inoculation	Bacteria: 0.1% of initial number by day 7 and further reduction or no increase thereafter (Spore-forming bacteria should remain static throughout the test period) Fungi: 10% of initial number by day 7 and further reduction (aqueous eye–area) or no increase thereafter

ASTM	*S. aureus* (6538) *Enterobacter aerogenes* (13048) *P. aeruginosa* (9027) *C. albicans* (10231) *A. niger* (16404) *Penicillium levitum* (10464) Cosmetic spoilage organisms	Mixed culture method (pure culture method optional) Rechallenge is recommended if product is subjected to repeated consumer insult (rechallenge should be at the 28th day of the initial challenge and continue test for additional 28 days)	Test at 0, 7, 14, 21, and 28 days (additional testing may be performed if appropriate)	Bacteria and yeasts: 0.1% of initial number by day 7 and no increase thereafter Other fungi: 10% within 28 days and no increase during the test period

of the challenge inoculum is also an important criterion in preservative tolerance, as is whether or not the cells are starved.[32a,32b] Greater resistance (i.e., slower in-product inactivation rates) has been described for broth-grown cultures vs. the same isolate grown on solid medium and suspended in saline.[33,34] The other consideration is use of inocula where the organisms may form spores when the desire is to measure killing of vegetative cells or vice versa. Inocula of fungi, for example, are typically composed of conidiospore suspensions obtained from sporulated agar cultures rather than mycelial forms.

Another source of inoculum, especially inoculum including adapted or tolerant microorganisms, is a similar, unpreserved or preserved product supporting a growing or active microbial population.[28] This inoculation vehicle offers one of the better tests of preservative efficacy. But it may not be suitable for use in testing of products significantly different from the contaminated vehicle.

Concentrations of microorganisms added to the test product are usually enough to achieve initial concentrations of about 1×10^6 CFU/gram for bacteria and 1×10^4 to 1×10^5 CFU/gram for fungi. Some manufacturers reportedly use even higher in-product microbial concentrations.[29] The only useful information this higher challenge provides is to identify the level of microbiological challenge that overwhelms the preservative system.[34]

The use of a mixed microbial culture inoculum vs. pure-culture inoculation of single, separate product samples also differs among current methods. Some consider a mixed inoculum to be more representative of actual conditions of contamination. It may also provide greater stress to the preservative system.[35] However, no published information supports this point of view.

A review of related technical observations indicates that either pure or mixed culture inocula have the potential to confound a preservative test. For example, the mixed-challenge inoculum introduces the dynamics of microbial species interactions into the challenge test. These factors can either amplify or diminish the degree of challenge offered the product. One factor that suggests mixed inocula are less resistant than single inocula is the ability of one organism to produce antibiotics effective against other consortial isolates.[36–38] Another factor is the competitive interaction among isolates for substrates and growth factors.[39] These may result in inhibition of organisms otherwise able to grow in the cosmetic product. In contrast, there are several interisolate factors that may favor consortial cooperation. These include cometabolism, microbial synthesis of vitamins or cofactors required by companion microorganisms, or the production of microbial surfactants and enzymes that directly or indirectly help growth.[40,41]

Henriette et. al.[42] described a viable mixed-bacterial population that developed in the presence of multiple disinfectants and antibiotics. Although none of the constituent microorganisms was resistant to all antimicrobial factors, the mixed population demonstrated growth and continued viability. To address such potential interactions, some methods suggest the use of mixed-inoculum subgroups composed of similar microorganisms. For example, one may have a group of Gram-positive bacteria and yeasts, a separate group of Gram-negative bacteria, and a third group composed of fungal conidiospores. It will be up to the microbiologist to determine the effective challenge provided by any inoculum, whether a single isolate or a mixed population. He or she should determine the effect of factors such as those described above.

Since multiple inoculation or rechallenge of a single product reportedly does not provide useful information on efficacy,[43] most methods specify only a single inoculation. However, some products such as eye makeup are commonly subjected to repeated insults by the consumer. These should be subjected to at least one rechallenge after the initial inoculum has been rendered nonrecoverable.

Conditions of Incubation and Inoculum Recovery

Following inoculation, one mixes the test product well and incubates it for at least 28 days. Incubation temperatures are typically 20 to 25°C under controlled conditions, since the efficacy of some cosmetic preservatives vary with temperature.[44] Samples of inoculated product are removed (from well-mixed product) at preselected intervals and diluted using an appropriate medium. The number of surviving microorganisms is then determined by plating aliquots of the diluted product.

Dilution of product samples serves two functions: preservative neutralization and, by serial dilution, precise estimation of surviving microorganisms. The more critical part of this operation is the neutralization of remaining product preservative. Effective chemical or physical means for neutralization of commonly used preservatives have been described.[45] These preservative-inactivating agents can be applied via the diluent or the plating medium, but typical practice uses the same agent in each step.

Two plating methods are specified for recovery and estimation of viable microorganisms: pour plating and spread or surface plating. In pour plates, diluted product is mixed with molten agar (44 to 48°C)[46] and dispensed into petri dishes. In spread plates, diluted product is spread on a solidified agar surface. Surface or spread plating allows easier processing of samples, avoids exposure of microorganisms to heated media, and may allow more aerobic incubation conditions. In contrast, pour plating may be more effective in diluting and neutralizing remnant levels of preservative.

Published information suggests that these two methods offer similar results but pour plating may be slightly superior to spread plating.[47,48] However, any application of recovery methods should be validated by appropriate methods.

During a PET, one must complete later sampling intervals despite apparent complete and rapid kill of the challenge inoculum at earlier intervals. Regrowth of the inoculum from apparently undetectable levels can occur late in the challenge test sampling sequence despite levels being less than detectable at earlier sampling times.[49] The mechanisms for this regrowth have not been characterized. Perhaps it is due to survival, adaptation, and subsequent growth from once undetectable levels of preservative-tolerant subpopulations. It may also be due to *in situ* recovery of injured organisms. Another mechanism may be repopulation of the product from adapting, container-associated organisms, or biofilms.[50,51,52,53] Unless one cultures the entire test volume at each sampling interval, one cannot be sure of complete kill of the inoculum. Since this is an impossibility, one follows a PET completely through 28 or more days of incubation.

The sensitivity of typical dilution and plate count methods in estimating numbers of bacterial survivors is low. It is usually from 10 to 20 microorganisms per gram of sample. The limits of detection depend on factors such as dilution and neutralization of the sample, and the volume of the sample that is cultured. In order to enhance sensitivity, Curry[52] proposed the use of broth enrichment of product samples. This method can detect low levels of viable microorganisms potentially undetectable by plate count methods. Here, product volumes greater than those used in plating procedures are diluted in broth and incubated. Enrichment increases the challenge test's sensitivity to detect surviving microorganisms. It is most useful in determining inoculum viability in samples from long-term tests showing no recovery in plate counts. However, the enrichment technique may not allow time for adaptation to show itself. Adaptation can occur in the form of a traditional Darwinian population drift with the development of individuals that are tolerant to the biocide.

Performance Criteria

The performance criteria that characterize an acceptably preserved product vary according to the compendial method used. Usually these criteria include a 99.9% or more decrease in bacterial viability within 1 to 2 weeks with continued reduction. For a fungal inoculum, the criteria include either no increase or up to a 90% reduction (of fungal viable elements) within 2 weeks. Using these test criteria, an acceptably preserved product would therefore affect only a minimal reduction of the bacterial inoculum. The bacteria would go down from 1×10^6 CFU/gm to 1×10^2 CFU/gm. The 100 CFU/gm result is consistent with the FDA-recommended, maximal final-product limit for nonpathogenic bacteria.[9] However, most, if not all, cosmetic manufacturers enforce far more stringent criteria of acceptance.[29] Sometimes this stringency goes down to a concentration less than the level of detection of most plate counting methods.[52]

METHODS IN USE

Long-term (28-day) Test Methods

A review of technical literature finds three primary methods proposed by industrial or compendial sources for challenge testing of cosmetics and related products. The sources are the Cosmetic Toiletry and Fragrance Association (CTFA methods), the American Society for Testing and Materials (ASTM method), and the U.S. Pharmacopeia (USP method). The Association for Official Analytical Chemists (AOAC) is pursuing a collaborative test method development with CTFA with heavy FDA involvement.

A comparison in Table 6-1 shows these methods to be similar in design and application. The primary differences between the methods are inoculum preparation (pure vs. mixed-culture) and the recovery of inoculated microorganisms (pour vs. spread plates). As discussed above, such procedural differences may be significant. Before a company commits to any method, including their own, they should support its use by appropriate validation. Validation includes demonstration of inoculum viability, its potential for preservative tolerance (if any), and adequate neutralization to permit recovery of the challenge microorganisms.

The methods in Table 6-1 also differ in what constitutes an acceptably preserved product. The USP method, though similar to other methods in bacterial standards, specifies only the prevention of an increase in viable fungal units (e.g., numbers of colonies). Apical growth of mycelial fungi does not always result in an increase in separate, viable units.[54] Therefore, satisfaction of the specification of no growth as measured by numbers of colonies does not mean that active fungal growth has stopped. The USP method nominally applies to "parenteral, otic, nasal, and ophthalmic products made with aqueous bases or vehicles" in their "original, unopened container." In this context and with adherence to good manufacturing practices, the issue of fungal growth may be met and thus the USP method may be appropriate for fungi.

CTFA conducted a survey among cosmetic manufacturers in 1986 (published in 1990) regarding the types of cosmetic preservative tests used.[29] Respondents included a total of 92 companies representing more than 200 products. The majority (78%) of the respondents was equally split between the use of in-house methods and modified CTFA procedures. The CTFA modifications included use of increased inoculum concentration, additional inoculum isolates, dilution of the product to be challenged, and more stringent acceptance criteria. A smaller percentage (19%) used the USP method and an even smaller group (3%) used the CTFA method as published. None used the ASTM method proposed for this purpose. Respondents also stated that their methods offered semiquantitative estimation of preservative efficacy in classifying a cosmetic product as poorly, marginally, or well-preserved for the purpose of determining if the product could withstand consumer use.

Most respondents in the survey reported that in-use tests confirmed the results of laboratory microbial challenge testing. Most manufacturers believed that their laboratory challenges provided a greater microbiological burden than in-use tests and therefore an added margin of safety. Thus, the study concluded that the in-use testing conducted by most cosmetic companies does mimic the microbial insults experienced by a product through the range of uses (and misuses) of consumer use.

Published methods[55,56] describe the design of these in-use tests. They typically place products in the hands of consumers who use product for a time. Following recovery from test panelists, the microbial content of each test product is estimated. Results from such studies show that products found to be well-preserved in laboratory challenge testing do not become contaminated in the consumers' hands.

Rapid Methods

Mulberry et al.[57] have reviewed the cosmetic microbiology methods available to simplify the screening for preservative efficacy of multiple formulations. These methods range from simple minimal inhibitory concentration (MIC) tests[35] to methods such as the linear regression-decimal reduction or D-value method,[58] the presumptive challenge method,[59] and the Accelerated Preservative Test (APT).[60]

A comparison of the latter three methods (Table 6-2) finds that its authors favorably compare the results to a long-term method. However, by virtue of the abbreviated timing implicit in their design, each of these methods can only observe the initial rate of microbiocidal activity. They are not useful in detecting potential tolerance development and regrowth of surviving challenge microorganisms from low numbers. Therefore, the presumptive challenge method and the APT are only screening tests and not replacements for longer-term tests such as the USP and CTFA methods.

In contrast, the author of the D-value method promotes it as a superior replacement for long-term challenge testing.[49] The basic assumption of the D-value method is that all members of a bacterial population are equally sensitive to a specific antimicrobial factor. With thermal death or gamma irradiation, this assumption is valid and such reactions are first order rate reactions. In cosmetics, the factor is the preserved product. The reactions here are second order requiring binding of the preservative to the organism first, then followed by the reaction that kills. With heat or radiation, the rate of kill is always linear through the point of extinction, also termed the "sterilization time."[61] However, in many systems including preserved products, nonlinear rates of kill occur.[62,63] Such observations warrant caution, especially for the complex formulas (providing multiple antimicrobial factors) and varied inocula used in cosmetic challenge testing. Caution related to these factors is probably responsible for the apparently limited acceptance of this method.[29]

Other Methods

Ten-Cycle, Multiple Challenge Method

The Dow Chemical Company ten-cycle challenge test has been proposed for a wide variety of products.[64] Dow developed the ten-cycle test, unlike the long-term tests or rapid tests, specifically to test those products that

TABLE 6-2

RAPID, SCREENING PRESERVATIVE CHALLENGE PROCEDURES FOR LIQUID COSMETICS

Method	Procedure	Sampling Intervals	Acceptance Criteria	Validation Method Cited
Presumptive Challenge	Product is serially diluted in saline suspension of challenge bacteria	24 hours following inoculation, loopful of each product dilution is streaked on a section of an agar plate	No established acceptance criteria, but product preservative efficacy may be compared to other products of known preservative efficacy	USP
Linear Regression—Decimal Reduction (D-Value)	Undiluted product is inoculated with microorganisms (to achieve in-product levels 10/ml for bacteria and 10/ml for fungi)	Surviving microorganisms estimated by plate counts 2 and 4 hours after inoculation for bacteria and 12 hours after inoculation for fungi	D-Value determined (time required to achieve 90% reduction in population viability). D-values of 4 hours or less for pathogenic bacteria, 28 hours or less for nonpathogenic bacteria and fungi are considered acceptable	USP, CTFA
Accelerated Preservative Test (ACP)	Product is diluted with glucose solution buffered to pH of product and challenged with microorganisms (to achieve levels in-product of 10/ml for bacteria and 10/ml for fungal spores	Surviving microorganisms estimated by plate counts at 0, 2, and 7 days	Reduction of bacterial numbers by 99% to 99.9% by day 2 and less than detection limit by day 7	Double challenge, consumer use

will undergo multiple contamination episodes during use. It is, therefore, a qualitative measure of a product's capacity to overcome multiple challenges.

The test consists of ten consecutive inoculations, each composed of 24-hour cultures of ATCC strains (one may use additional resistant strains). Each inoculation is separated by at least 24 hours from the previous inoculation event. Before each inoculation event, a sample is removed and streaked on the agar plate to obtain a semiquantitative estimate of microbial content.

In the Dow test, preservation of the personal care product is adequate if no microorganisms are recovered at any sampling interval or at the test's completion. No data are available that correlate results from the ten-cycle method with in-use testing or offer validation with other test methods.

Modeling

Food microbiologists now use mathematical models to predict the shelf life of foods. These mathematical models estimate the combined effects of pH, temperature, water activity, carbonation, acidulant, and preservative concentrations on food spoilage.[65–67] Some of the models[66] include those that describe the chances of a microbial event occurring (probabilistic), those which project changes in bacterial numbers (kinetic), and those that predict the timing of a particular microbiological event ("response surface").

Some scientists criticize such models[68] since they are descriptive rather than predictive. However, the potential value of such systems justifies modeling as a major research initiative in both the U.K. and the U.S. If predictive in food applications, modeling systems might also be useful to the cosmetic microbiologist. Models may systematize the effects of multiple formulation factors. They may be especially useful for optimizing synergistic combinations of antimicrobials and compounds or factors that help preservative activity.[61,69–72]

SUMMARY

We need effective preservation of cosmetic products. The negative effects of contamination on product quality, regulatory compliance, and on consumer health are great. Simple addition of chemical preservatives is not enough to insure product quality. Manufacturers must confirm antimicrobial efficacy through appropriate testing. The purpose of such tests is not to prove absolute preservation efficacy under all conditions and vs. all microorganisms. Rather, it is a test that estimates the efficacy of preservation within the parameters anticipated for the proposed product application.

Preservation is only a part of a product's microbiological quality. It should not be expected to correct massive contamination that might result from improper manufacturing practices or contaminated raw materials.[73] Instead, preservation reduces the limited risk of contamination that exists despite the use of acceptable manufacturing practices. It also reduces the anticipated risk of contamination from consumer use and misuse.

Accordingly, one should validate any method used to estimate preservative efficacy. Validation should, at minimum, insure that the exposure of a representative product (formula and package) to an appropriate microbial challenge is possible without the product becoming contaminated. It should demonstrate effective recovery of any viable microorganisms through the entire incubation period. Where possible, challenge testing should be supported by appropriate chemical analysis of preservative stability. Acceptable antimicrobial efficacy should be determined vs. standards appropriate for the product type, anticipating the product's shelf life and consumer use and misuse practices. Finally, the results of laboratory testing methods should be validated through in-use experiments.

REFERENCES

1. 75th Congress. (1983). *Federal Food Drug and Cosmetic Act*, and as later amended, 21 U.S. Code, 201(i): p.2.
2. Geis, P. (1988). Preservation of cosmetics and consumer products: rationale and application, *Dev. Indust. Microbiol.*, 29, 305.
3. Morse, L. J., Williams, H. L., Green, F. P., Eldridge, E. E., and Rotta, J. R. (1965). Septicemia due to *Klebsiella pneumoniae* originating from a hand-cream dispenser, *N. Engl. J. Med.*, 277, 472.
4. O'Day, D. M., Head, W. S., and Robinson, R. D. (1987). An outbreak of *Candida parapsilosis* endophthalmitis: an analysis of strains by enzyme profile and antifungal susceptibility, *Br. J. Ophthalmol.*, 71, 126.
5. Reis, F. R. and Wood, T. O. (1979). *Pseudomonas* corneal ulcer: the causative role of eye cosmetics, *Arch. Ophthalmol.*, 97, 1640.
6. Wilson, L. A. and Ahearn, D. G. (1977). *Pseudomonas*-associated corneal ulcer associated with contaminated eye mascaras, *Am. J. Ophthalmol.*, 84, 112.
7. Fainstein, V., Andres, N., Umphrey, J., and Hopfer, R. (1988). Hair clipping: another hazard for granulocytopenic patients, *J. Infect. Dis.*, 158, 655.
8. Eierman, H. J. (1984). Cosmetic product preservation: safety and regulatory issues, *Cosmetic and Drug Preservation: Principles and Practices* (J. J. Kabara, Ed.), Marcel Dekker, Inc., New York, p. 559.
9. Madden, J. M. and Jackson, G. J. (1981). Cosmetic preservation and microbes: viewpoint of the Food and Drug Administration, *Cosmet. Toilet.*, 96, 75.
10. Preservation Subcommittee: Cosmetic Toiletry and Fragrance Association Microbiology Committee. (1985). Determining the need for preservatives in cosmetics and toiletries (weighted value technique), *CTFA Microtopics* 1988, 83.
11. United States Pharmacopeia Convention. (1995). Microbiological tests, antimicrobial preservatives-effectiveness. *U.S. Pharmacopeia XXIII*, U.S. Pharmacopeial Convention, Rockford, Maryland.

12. Schneider, F.H. (1984). Evaluation of chemical toxicology of cosmetics, *Cosmetic and Drug Preservation: Principles and Practices* (J. J. Kabara, Ed.), Marcel Dekker, Inc., New York, p. 533.
13. Bassett, D. J. C. (1971). Common source outbreaks, *Proc. Royal Soc. Med.*, 64, 980.
14. Anderson, R. L., Berkelman, R. L., Mackel, D. C., Davis, B. J., Holland, B. W., and Martone, W. J. (1984). Investigations into the survival of *Pseudomonas aeruginosa* in polaxmer-iodine, *Appl. Environ. Microbiol.*, 47, 757.
15. Ansell, J. M. (1984). Contamination of germicidal and nongermicidal agents, *Am. Clin. Pathol. Rev.*, 1984, 32.
16. Brannan, D. K. and Dille, J. C. (1990). Type of closure prevents microbial contamination of cosmetics during consumer use, *Appl. Environ. Microbiol.*, 56, 1476.
17. CTFA Preservation Subcommittee. (1981). A guideline for the determination of adequacy of preservation of cosmetic and toiletry formulations, in *CTFA Guidelines*, Cosmetic, Toiletry, and Fragrance Association, Washington, D.C.
18. CTFA Preservation Subcommittee. (1981). A guide for preservative testing of aqueous and semi-aqueous liquid and semi-liquid eye area cosmetics, in *CTFA Technical Guidelines*, Cosmetic, Toiletry, and Fragrance Association, Washington, D.C.
19. American Society for Testing and Materials. (1991). Standard method for water-containing cosmetics (ANSI/ASTM E 640-78), *Annual Book of ASTM Standards*, volume 11.04, American Society for Testing Materials, Philadelphia, p.290.
20. Anonymous. (1990). Cosmetic preservatives encyclopedia. Antimicrobials, *Cosmet. Toilet.*, 105, 49.
21. Wallhauser, K. H. (1984). Antimicrobial preservatives used by the cosmetic industry, *Cosmetic and Drug Preservation*, J. J. Kabara, Ed., Marcel Dekker, Inc., New York, p.605.
22. Benassi, C. A., Rettore, A., Semenzato, A., Bettero, A., and Cerini, R. (1988). Chemical stability and microbiological reply of preservative systems in cosmetics: a direct correlation, *Int. J. Cosmet. Sci.*, 210, 2312.
23. Semenzato, A., Benassi, C. A., Rossi, G., Bettero, A., Lucchari, M., and Cerini, R. (1990). Dowicil 200 stability test: chemical and microbiological studies, *Int. J. Cosmet. Sci.*, 12, 265.
24. Armstrong, N. A. (1972). Uptake of preservatives by plastic packaging, *Am. Cosmet. Perfum.*, 87, 45.
25. Power, E. G. M. and Russell, A. D. (1989). Glutaraldehyde: its uptake by sporing and non-sporing bacteria, rubber, plastic, and an endospore, *J. Appl. Bacteriol.*, 67, 329.
26. Anderson, R. L., Vess, R. W., Panlilio, A. L., and Favero, M. S. (1990). Prolonged survival of *Pseudomonas cepacia* in commercially-manufactured povidone- iodine, *Appl. Environ. Microbiol.*, 56, 3598.
27. Hugo, W. B., Pallent, L. J., Grant, D. W., Denyer, S. P., and Davies, A. (1986). Factors contributing to the survival of a strain of *Pseudomonas cepacia* in chlorhexidine solutions, *Lett. Appl. Microbiol.*, 2, 37.
28. Cooke, P. K., Gandhi, U. R., Lashen, E. S., and Leasure, E. L. (1991). Preservative evaluation: designing an improved test system, *J. Coat. Technol.*, 63, 33.

29. Microbiology Committee: Cosmetic Toiletry and Fragrance Association. (1990). CTFA survey: test methods companies use, *Cosmet. Toilet.*, 106, 79.
30. Levy, E. (1987). Insights into microbial adaptation to cosmetics and pharmaceutical products, *Cosmet. Toilet.*, 102, 69.
31. Favet, J., Fehr, A., Griffiths, W., Amacker, P. A., and Schorer, E. (1987). Adaptation of *Escherichia coli, Pseudomonas aeruginosa*, and *Staphylococcus aureus* to Kathon CG and Germal II in an O/W cream, *Cosmet. Toilet.*, 102, 75.
32. Orth, D. S. and Lutes, C. M. (1985). Adaptation of bacteria to cosmetic preservatives, *Cosmet. Toilet.*, 100, 57.
32a. Brown, M. R. W. and Williams, P. (1985). Influence of substrate limitation and growth phase on sensitivity to antimicrobial agents. *J. Antimicrob. Chemother.* 15, Suppl. A, 7.
32b. Gilbert, P., Collier, P. J., and Brown, M. R. W. (1990). Influence of growth rate on susceptibility to antimicrobial agents: biofilms, cell cycle, dormancy, and stringent response. *Antimicrob. Ag. Chemother.*, 34, 1865.
33. Al-Hiti, M. M. and Gilbert, P. (1983). A note on inoculum reproducibility: a comparison between solid and liquid culture, *J. Appl. Microbiol.*, 55, 173.
34. Orth, D. S., Lutes, C. M., and Smith, D. K. (1989). Effect of culture condition and method of inoculum preparation on the kinetics of bacterial death during preservative efficacy testing, *J. Soc. Cosmet. Chem.*, 40, 193.
35. Parsons, T. (1990). A microbiological primer for the microbiology manager, *Cosmet. Toilet.*, 105, 73.
36. Lim, H-S. and Kim, S-D. (1991). *Pseudomonas stutzeri* YPL-1 genetic transformation and antifungal mechanism against *Fusarium solani*, an agent of plant root rot, *Appl. Environ. Microbiol.*, 578, 510.
37. McKay, A. M. (1990). Antimicrobial activity of *Enteroccocus faecium* against *Listeria* spp., *Lett. Appl. Microbiol.*, 11, 15.
38. Harrison, L., Teplow, D. B., Rinaldi, M., and Strobel, G. (1991). Pseudomycins, a family of peptides from *Pseudomonas syringae* possessing broad-spectrum antifungal activity, *J. Gen. Microbiol.*, 137, 2857.
39. Neilands, J. B. (1981). Microbial iron compounds, *Ann. Rev. Biochem.*, 50, 715.
40. Knudsen, G. R., Eschen, D. J., Danduranc, L. M., and Wang, Z. G. (1991). Methods to enhance growth and sporulation of pelletized biocontrol fungi, *Appl. Environ. Microbiol.*, 57, 2864.
41. Sondossi, M., Rossmore, H. W., and Wireman, J. W. (1986). Induction and selection of formaldehyde-based resistance in *Pseudomonas aeruginosa*, *J. Indust. Microbiol.*, 6, 97.
42. Henriette, C., Petitdmange, E., Ravel, G., and Gray, R. (1991). Mercuric reductase activity in the adaptation to cationic mercury, phenyl mercuric acetate and mixed antibiotics of a Gram-negative population isolated from an aerobic fixed-bed reactor, *J. Appl. Microbiol.*, 71, 439.
43. Preservation Subcommittee: Cosmetic Toiletry and Fragrance Association Microbiology Subcommittee. (1981). A study of the use of rechallenge in preservation testing of cosmetics, *CTFA Cosmet. J.*, 13, 19.
44. Gililand, D., Li Wan Po, A., and Scott, E. (1991). The bactericidal activity of a methyl and propyl parabens combination isothermal and non-isothermal studies, *J. Appl. Bacteriol.*, 72, 252.

45. Singer, S. (1987). The use of preservative neutralizers in diluents and plating media, *Cosmet. Toilet.*, 102, 55.
46. CTFA Methodology Subcommittee (1985). Determination of the microbial content of cosmetic products, in *CTFA Technical Guidelines*, Cosmetic, Toiletry, and Fragrance Association, Washington, D.C.
47. Beerens, H., Romond, C., and Lemarie, D. (1976). Techniques for the enumeration of microorganisms in cosmetic products, *Soaps Perfum. Cosmet.*, 49, 262.
48. El-Zanfaly, H. T. and Sabbour, M. M. (1984). Evaluation of pour and surface plate methods for *Staphylococcus aureus* enumeration, *Arch. Lebensm. Hyg.*, 34, 133.
49. Orth, D. S. (1991). Standardizing preservative efficacy test data, *Cosmet. Toilet.*, 106, 45.
50. Russell, A. D. (1991). Injured bacteria: occurrence and possible significance, *Lett. Appl. Microbiol.*, 12, 1.
51. van Loosdrecht, M. C. M., Lyklema, J., Norde, W., and Zehnder, A. J. B. (1990). Influence of interfaces on microbial activity, *Microbiol. Rev.*, 54, 75.
52. Curry, J. (1987). Thoughts on preservation testing of water-based products, *Cosmet. Toilet.*, 102, 93.
53. Delaquis, P. J., Caldwell, D. E., Lawrence, J. R., and McCurdy, A. R. (1989). Detachment of *Pseudomonas fluorescens* from biofilms on glass surfaces in response to nutrient stress, *Micro. Ecol.*, 18, 1999.
54. Moore-Landecker, E. (1972). *Fundamentals of the Fungi*, Prentice Hall, Inc., Englewood Cliffs, New Jersey, p. 7.
55. Brannan, D. K., Dille, J. C., and Kaufman, D. J. (1987). Correlation of in vitro challenge with consumer-use testing for cosmetic products, *Appl. Environ. Microbiol.*, 53, 1827.
56. Lindstrom, S. M. (1986). Consumer use testing: assurance of microbiological product safety, *Cosmet. Toilet.*, 101, 712.
57. Mulberry, G. K., Entrup, M. R., and Agin, J. R. (1987). Rapid screening methods for preservative efficacy evaluations, *Cosmet. Toilet.*, 102, 47.
58. Orth, D. S. (1980). Establishing cosmetic preservatives by use of D-values, *J. Soc. Cosmet. Chem.*, 31, 165.
59. Chan, M. and Prince, H. (1981). Rapid screening test for ranking preservative efficacy, *Drug. Cosmet. Ind.*, 129, 34.
60. Scibienski, E. J., O'Neill, J. J., and Mead, C. A. (1981). An accelerated preservative test, *Cosmet. Toilet.*, 96, 91 (abstract).
61. Orth, D. S., Lutes-Anderson, C. M., Smith, D. K., and Milstein, S. R. (1989). Synergism of preservative system components: use of survival curve slope method to demonstrate anti-*Pseudomonas* synergy of methyl paraben and acrylic acid homopolymer/copolymers *in vitro*, *J. Soc. Cosmet. Chem.*, 40, 347.
62. Cerf, O. (1989). Tailing of survival curves of bacterial spores, *J. Appl. Bacteriol.*, 42, 1.
63. Campbell, J. E. and Dimmick, K. L. (1966). Effect of 3% hydrogen peroxide on the viability of *Serratia marcescens*, *J. Bacteriol.*, 91, 925.
64. Sabourin, J. R., (1990). Evaluation of preservatives for cosmetic products, *Drug Cosmet. Ind.*, 6, 24.

65. Adams, M. R., Little, C. L., and Easter, M. C. (1991). Modelling the effect of pH, Acidulant and temperature on the growth rate of *Yersinia entercolitica*, *J. Appl. Microbiol.*, 71, 65.
66. Cole, M. (1991). Predictive modeling--yes it is!, *Lett. Appl. Microbiol.*, 13, 218.
67. Graham, G. (1991). Predictive mathematical modeling of microbial growth and survival in foods, *Food Sci. Technol. Today*, 3, 2.
68. Hedges, A. (1991). Predictive modeling—or is it?, *Lett. Appl. Microbiol.*, 13, 217.
69. Anon. (1990). Compounds that contribute to preservative activity, *Cosmet. Toilet.*, 105, 63.
70. Russell, A. D. (1990). Antimicrobial activity associated with non-antimicrobial agents, *Lett. Appl. Microbiol.*, 11, 171.
71. Whaley, G. (1991). Preservative properties of EDTA, *Manuf. Chem.*, 62, 22.
72. Gililand, D., Li Wan Po, A., and Scott, E. (1991). Kinectic evaluation of claimed synergistic paraben combinations using a factorial design, *J. Appl. Bacteriol.*, 72, 258.
73. Bryce, D. M. (1990). Quality in a commercial environment, *Cosmet. Toilet.*, 105, 83.

7 MECHANISMS OF ACTION OF COSMETIC PRESERVATIVES

Thomas E. Sox

CONTENTS

INTRODUCTION: BACKGROUND ON MECHANISMS OF ACTION OF PRESERVATIVES

The cosmetics industry uses a wide range of chemical agents to prevent microbial growth in cosmetic products. Development or discovery of the older agents occurred through a process of trial and error. The cosmetics industry has used these agents for decades without a detailed understanding of their mechanisms of action.

It is not necessary to understand the mechanism of action of an antimicrobial agent to successfully use it in product preservation. However, there are many occasions in which an understanding of the mechanism of action can benefit the formulator or microbiologist. Knowing the mechanisms of action may permit the prediction of whether the combination of two antimicrobial agents will be synergistic or antagonistic. For example, an agent that increases the permeability of the cell may produce an increase in the efficacy of an agent that affects the cytoplasm. Similar situations exist with antibiotics. The concurrent exposure of microorganisms to two antibiotics can lead to bona fide synergism (e.g., amoxicillin and clavulinic acid).

0-8493-3713-5/97/$0.00+$.50

Knowing a compound's chemistry and mechanism of action also aids in the intelligent selection of neutralizers to incorporate into the recovery media. The microbiologist should develop a neutralizer for each preservative under evaluation. This allows laboratory methods to distinguish between the microbiostatic and microbiocidal actions of the preservative. The selection of potential neutralizers is simpler if one knows the antimicrobial mode of action and the chemical properties of the preservative.

Finally, an understanding of the chemistry and mechanism of action of the biocide helps in predicting interactions between the biocide and other product components. This interaction may adversely affect the antimicrobial efficacy. Or, the interaction may adversely affect the other product ingredients.

MECHANISMS OF ACTION OF MAJOR COSMETIC PRESERVATIVES

This chapter reviews the available data on mechanisms of action of major cosmetic preservatives used in the U.S. Some of the preservatives are grouped together, like the commonly used esters of para-hydroxybenzoic acid, since these are likely to have similar mechanisms of action. Therefore, the discussion will center on the following functional categories of cosmetic preservatives: paraben esters, phenol derivatives, alcohols, organic acids and salts, isothiazolinone compounds, formaldehyde donors, bromonitro compounds, and chelating agents.

Frequently, there is only limited data available for some preservative groups. The true mechanism of action for some of these remains unknown. In the following sections, we review available information on the mechanisms of action of these eight major groups of antimicrobial preservatives. These eight groups are not all the antimicrobial preservatives now used in cosmetic products in the U.S. However, it does encompass the commonly used preservatives as disclosed in a recent FDA survey.[1] We will not cover synergistic effects resulting from the combination of antimicrobial preservatives. Instead, Denyer, Hugo, and Harding[2] provide a thorough review of this topic.

Paraben Esters

Parabens (paraben esters or p-hydroxybenzoate esters) are the most widely used preservatives in cosmetic products. The methyl, *n*-propyl, and *n*-butyl esters are the predominant forms used.[1] A wide range of suppliers provide these materials under diverse trade names. The paraben esters come either singly or in combination. Some suppliers combine one or more parabens with other antimicrobial preservatives. These combinations simplify the formulation of products, since it decreases the number of raw materials involved in compounding.

One drawback of using a prepared mixture of parabens and other preservatives is that one of the preservatives may be incompatible for a given formula. In order to prove that the preservatives in a combination make a contribution to product preservation, the preservative efficacy test should use suitable control products containing only the individual preservatives. This way, one can show that each of the preservatives in a combination are making a contribution to product preservation. An alternative hypothesis to explain these results is disruption of the chemiosmotic energy potential for ATP generation.

There is extensive literature on the antimicrobial activity of the paraben esters against a wide variety of microorganisms. There is much less work on the mode of action of these compounds. Separating antimicrobial effects of parabens from secondary physiological effects of paraben-exposed microorganisms has been a key challenge for research in this area. Determining the primary effects is essential for identifying the target site of the compound.

Furr and Russell[3] investigated the effects of paraben esters on *Serratia marcescens* growth and membrane permeability. They measured changes in membrane permeability by decreases in optical density of exposed cells and the leakage of pentoses from the cells.

Cells exposed to parabens leaked intracellular contents. The investigators saw no overt change in cell structure in electron micrographs of exposed cells. So the leakage was not the result of cell lysis. Magnesium ions, which stabilize bacterial membranes,[4] decreased the extent of leakage in *S. marcescens* exposed to parabens.

Leakage was greater after exposure to propyl or butyl paraben than after exposure to methyl or ethyl parabens. This observation is consistent with the greater effect that propyl and butyl esters have on the viability of washed cell suspensions. Detectable leakage of cellular components occurred at subinhibitory levels of parabens. This showed that cells were able to survive some degree of membrane damage and loss of intracellular components.

Freese et al.[5] noted that parabens and lipophilic acids used as preservatives inhibit the uptake of serine in membrane vesicles of *Bacillus subtilis.* This inhibition was concentration dependent and similar to that seen for intact cells being inhibited by the same compound. The inhibitory activity of the parabens on serine uptake and growth increased as the chain length of the parabens increased. For example, growth inhibition by butyl paraben was more than ten times that produced by an equimolar concentration of methyl paraben. However, these results did not conclusively show that transport inhibition was the actual mode of action of these compounds. Instead, the inhibition of transport may reflect the disruption of other essential cellular processes by these compounds.

Freese et al. also noted that the growth inhibitory activity of propyl paraben was reversible.[5] Cells inhibited by exposure to propyl paraben in the medium resumed growth at the uninhibited rates immediately after transfer to fresh medium without propyl paraben. This suggests that inhibition by parabens is not the result of irreversible chemical reactions between the compound and cell components. Instead, inhibition results from an uptake of the compound by one or more cell structures. This uptake is concentration dependent and readily reversible if the extracellular concentration is decreased.

The chemiosmotic hypothesis has profoundly affected the understanding of cell energetics. According to the chemiosmotic hypothesis,[6] viable cells maintain an energy potential across the cell membrane composed of two parts. One component is a pH differential between the interior of the cell and the external environment. The interior of the cell is more alkaline than the environment. The pH differential occurs because of the energy-dependent transport of protons out of the cell. The second component is the electrical charge across the cell membrane. The interior of the cell maintains a negative electrical charge compared to the exterior. This energy potential across the membrane is crucial to many cell processes. These processes include transport of substrates into the cell and the generation of ATP to provide energy for intracellular processes.

Eklund[7] found that methyl and butyl parabens at concentrations of 0.01 m*M* to 0.1 m*M* disrupted the membrane pH gradient in membrane vescicles of *Escherichia coli.* However, they needed about 10 times this concentration to have a significant impact on the electrical charge potential across the membrane. These concentrations were about the same as those needed to prevent growth of whole cells. Thus, their work suggested that disruption of the membrane electrical potential is the actual mechanism by which parabens prevent microbial growth.

It appears that effects such as the decreased transport of amino acids[5] may be secondary effects resulting from the decreased ability of paraben-exposed cells to carry out energy-requiring processes such as active transport. Thus parabens appear to fall into the first category established by Davidson and Branen:[8] compounds that decrease control of membrane permeability. Additional work should be done to determine the extent and nature of paraben incorporation into membranes.

Phenol Derivatives

Cosmetic products only infrequently have phenol derivatives as preservatives.[1] The phenols most widely used in cosmetics are ortho-phenylphenol and its sodium salt, chloroxylenol, and resorcinol. Phenol retains some use as a disinfectant. It receives little use as a cosmetic preservative because of its toxicity to humans. The cosmetic industry more widely uses ring-substituted derivatives of phenol since they have increased antimicrobial activity.

The antimicrobial activities of phenol derivatives vary widely with the nature and location of the substituent groups. Extensive investigation into the relationships between structure and antimicrobial activity of these compounds was done early in the 20th century. Suter reviewed the results of this work in 1941.[9]

The degree of partitioning of the compound between oil and water is a key parameter in determining the antimicrobial activity of the phenolic compound.[10] More lipophilic compounds have greater antibacterial activity. This may be due to the tendency of these materials to partition into lipid-containing bacterial membranes. Prindle reviewed the antimicrobial activity of various phenol derivatives.[10] Phenol derivatives appear to function at low concentrations by disrupting the proton motive force,[11,12] suggesting the cell membrane is the target site. Thus, phenols may have a mechanism of action similar to that of the paraben esters. The paraben esters are somewhat similar structurally to phenol derivatives. Therefore, phenol derivatives, like parabens, fall into the first mechanism category of Davidson and Branen.[8]

Alcohols

Several alcohols are useful as preservatives in cosmetic products. The concentration needed to achieve effective preservation varies widely with the alcohol and the specific application. Food, pharmaceutical, and drug products use ethanol (ethyl alcohol) widely as a preservative. This use requires high concentrations (usually greater than 4% of the total formula weight) of ethanol to prevent growth of bacteria and fungi. At levels greater than 20%, the product is considered completely "self-sterilizing." The high levels required prevent its use in many applications.

Propylene glycol has also received some use as a preservative and has the additional benefit of exerting a synergistic effect with paraben esters. Propylene glycol again is only effective in quite high concentrations and may function via the same mode of action as ethanol. Phenoxyethanol and benzyl alcohol are the most widely used preservative alcohols in cosmetic products.[1] Another cosmetic preservative in use is phenethyl alcohol. The mode of action of this aromatic alcohol has received the most attention. The structurally similar phenoxyethanol and benzyl alcohol may share this mechanism of action.

Phenethyl alcohol appears to exert its antimicrobial activity via effects on membrane permeability. Membrane permeability of *E. coli* increased after exposure to phenethyl alcohol (PEA).[13] An efflux of intracellular potassium ions suggested the loss of membrane permeability control. Acriflavine, a dye to which cells are normally impermeable, penetrated the cells after PEA exposure. The activity of PEA in disrupting the membrane permeability barrier was very similar to the effects of toluene. Toluene is routinely used to increase the permeability of cells in bacterial physiology experiments. It allows the entry of materials that are normally unable to penetrate the cell.[13]

Phenethyl alcohol may also inhibit DNA synthesis.[14] This inhibition may be direct, since DNA synthesis occurs in a membrane-associated complex. Alternatively, it may be a secondary effect that is a sequel to massive disruption of intracellular functions. This multiple site of action effect by such agents is typical of the difficulty in discerning between primary and secondary effects of broad-spectrum antimicrobial compounds.

The antimicrobial action of phenoxyethanol has also received attention. Hall summarized the work done on this agent in 1984.[15] This work showed there were several different effects of phenoxyethanol against bacteria. The compound was inhibitory against enzymes involved in the tricarboxylic acid pathway.[16] Also, bactericidal levels of phenoxyethanol caused leakage of intracellular materials and a loss of permeability control.[17] Levels of the compound below the minimum inhibitory concentration caused changes in membrane permeability and outer membrane structure in *E. coli.* These latter effects may be significant for evaluating preservative activity since they occur at lower concentrations than is required to severely damage the exposed cells and halt their growth.

In summary, these compounds appear to function as a result of disrupting cell membrane permeability. But, inhibition of essential enzymes, the second category of Davidson and Branen,[8] may also be involved.

Organic Acids and Salts

Several weak acids are useful in the preservation of acidic products. These are typically effective only in formulations with pH levels below 5. The most widely used members of this category are benzoic acid and its sodium salt, sorbic acid and its potassium salt, and dehydroacetic acid and its sodium salt.[1] One usually formulates these materials into products in either the acid form or as the sodium and potassium salts. One frequently prefers the salts because of their greater aqueous solubility. However, the formulator must be aware that the addition of these salts can affect the pH of weakly buffered acidic products.

The sensitivities of individual microorganisms to each of these compounds is similar. This suggests that they may act through similar mechanisms.[18] Organic acids are effective only under acidic conditions because only the protonated forms, which exist at the acidic pH, can pass through the cell membrane. The uncharged nature of the protonated form allows penetration of the hydrophobic lipid membrane. In contrast, the anionic forms that predominate at higher pH levels cannot cross the cell membrane because of their charged nature. One uses these and several other organic acids, particularly propionic acid and lactic acid and their sodium salts, as preservatives in acidic foods. As a result, much of the available information on the mode of action of these compounds is in the food technology literature.

Weak acids penetrate cells under acidic conditions. These acid preservatives exist in a protonated state in the acidic extracellular environment. The hypothesis for their mode of action is that they pass into the less acidic cytoplasm of the cell, the acid dissociates, releasing a proton (or hydrogen ion) and thus decreasing the intracellular pH. The active transport process that maintains the proton gradient across the cell membrane must therefore pump additional protons out of the cell to maintain a certain energy efficiency. Since this requires more energy, the cell is slowly starved to death.

Warth used yeast, for example, to show that the intracellular concentrations of preservative did not reach the equilibrium levels predicted from their extracellular concentration and the intracellular pH.[18] Exposure of cells to these preservatives caused an increase in respiration that was not coupled with an increase in growth. Warth postulated that a transport process to remove the weak acid from the cell may exist which consumed any energy generated.[19]

Isothiazolinone Compounds

Isothiazolinone compounds have received increasing use as preservatives in cosmetic products over the last decade. In many applications, very low levels of these compounds provide adequate preservation. This factor, in combination with efficacy against a broad range of microorganisms, is responsible for their widespread use. A mixture of 5-chloro-2-methyl-3(2H)-isothiazolone (CMIT) and 2-methyl-3(2H)-isothiazolone (MIT) is sold as Kathon CG (a registered trademark of the Rohm and Haas Company) for preservation of cosmetic products. Information on the antimicrobial spectrum and preservative efficacy of this material has been available for some time.[20,21]

Within the past several years, information on the mode of action of isothiazolones was published.[22–24] In earlier work, a similar isothiazolinone compound, 1,2-benzisothiazolinone (BIT), inhibited the active transport and glucose oxidation of *Staphylococcus aureus* NCTC 6571. Glucose oxidation was inhibited by 94% at a BIT concentration equivalent to the Minimum Inhibitory Concentration (MIC) in nutrient broth for this strain.[21] Under these conditions, BIT inhibited active transport of glucose by 70%. Since BIT caused only a slight leakage of potassium from the cells, the authors concluded that it did not significantly affect membrane integrity.

The addition of thiol compounds to the growth medium increased the MIC against *S. aureus.* This result suggested that BIT reacted with cellular thiol groups. The demonstration that BIT inhibited isolated thiol-containing enzymes like ATPase and glyceraldehyde-3-phosphate dehydrogenase continued to support this hypothesis. It had no effect on asparaginase, which lacked thiol groups.[21]

More recently, Collier et al.[22] used ^{14}C-radiolabeled CMIT and MIT to evaluate uptake and partitioning of these compounds by *E. coli* ATCC 8739 and the yeast *Schizosaccharomyces pombe* NCYC 1354. Uptake by both organisms was rapid, with equilibrium occurring within 3 minutes. The absorption patterns by the yeast resembled C-type adsorption isotherms[25] indicative of uptake that is directly proportional to the biocide concentration, and general partitioning throughout the contents of the cell. In contrast, S-type isotherms occurred with *E. coli* exposed to MIT, while C-type isotherms occurred for the same strain exposed to CMIT. The isotherm for MIT suggests facilitated uptake, in which the binding of biocide molecules would enhance the subsequent binding of other biocide molecules.

Isothiazolone biocides undergo several reactions with cell components after binding. The interaction of isothiazolone biocides with microbial cells proceeds via several sequential reactions.[23] Initially, a biocide molecule reacts with a thiol group to form a disulfide linkage with the biocide. This structure then interacts with a second thiol group to establish a new disulfide linkage. Following this, the cell releases the biocide as a mercaptoacrylamide.

The mercaptoacrylamide derivative of CMIT tautomerizes to a thioacyl chloride under physiological conditions. The thioacyl chloride is highly reactive and proceeds to interact with other cell macromolecules. The thioacyl chloride does not react just with thiol groups. Collier et al. incubated isothiazolone biocides with or without dithiothreitol before exposure to ethanol dehydrogenase. This enzyme lacks a readily attacked thiol group. Pre-incubation with dithiothreitol did not increase the inhibition of ethanol dehydrogenase by BIT or MIT, but inhibition by CMIT increased by exposure to dithiothreitol. This inhibition was due to the generation of the thioacyl chloride derivative through reaction of CMIT with dithiothreitol.

The ability of CMIT to degrade to a thioacyl chloride may account for its much greater antimicrobial activity compared to BIT or MIT. In tests with *E. coli* ATCC 8739, CMIT had an MIC about 100-fold less than that of MIT and 30-fold less than that to BIT.[24] This novel reactivity of CMIT may account for some of its unique toxicological properties.[26] Also, Collier et al.[23] speculated that the thioacyl chloride may react with nucleic acids. This could account for the positive response that CMIT produces in the Ames test.

Inhibition of intact cells by CMIT does not appear to proceed by interaction with specific cell proteins. Analysis of proteins from ^{14}C-CMIT exposed cells did not indicate any preferential binding of biocide to specific proteins.[22] Hence, the antimicrobial activity of this compound may lie in its nonspecific reactivity with thiol-containing enzymes in the cell envelope. Or, the activity may be due to subsequent reactions of the thioacyl chloride with other cell components.

Although recent work on the mode of action of isothiazolinones has advanced our understanding of the materials, it is unclear into which category of Davidson and Branen[8] to place the isothiazolinones. Available data implicate damage to essential enzyme systems, but effects on membrane permeability and nucleic acids may also occur. Clearly the isothiazolinones attack at multiple sites with multiple modes of action. This feature may help limit resistant strain development even though some resistant strains may develop based on exclusion or on extracellular glutathione production.

Formaldehyde Donors

Formaldehyde has a long history of use as an antimicrobial agent. Aqueous formaldehyde solutions as cosmetic preservatives are rarely used today. Instead, we now use organic compounds that gradually hydrolyze in aqueous products to yield microbiocidal levels of formaldehyde. The extent of formaldehyde release varies depending on the various compounds.[27]

The most widely used formaldehyde releasers are dimethyloldimethyl hydantoin (Gyldant, Glyco, Inc.), imidazolidinyl urea derivatives (Germall 115 and Germall II, Sutton Laboratories, Inc.), polymethoxy bicyclic oxazolidine (Nuosept C, Hüls), and 1-(3-chloroallyl)-3,5,7-triaza-1-azoniaadamantane hydrochloride (Quaternium 15 or Dowicil 200, Dow Chemical Co.). These compounds apparently differ in the amounts of free formaldehyde released into solution.[27] It is likely that the actual antimicrobial compound is the released formaldehyde either as it is released into the formula or as it is released into the microbe.

Formaldehyde has the potential to react with several sites of biological importance. It can alkylate amino and sulfhydryl groups of proteins as well as the ring nitrogens of purine bases.[28] This high reactivity probably accounts for the extremely broad antimicrobial spectra of both formaldehyde itself and the formaldehyde donor compounds.

Very little has been done to determine the exact sites of reactivity for formaldehyde in intact cells. Pursuit of additional work in this area might provide a more detailed understanding of the exact antimicrobial mode of action of formaldehyde. Exposure of intact cells to cidal concentrations of radiolabeled formaldehyde and subsequent analysis of the distribution of formaldehyde among the cell macromolecules would identify the specific interactions that take place in the exposed cells. This work alone would not pinpoint the exact interaction responsible for cell death, however. Formaldehyde is highly reactive and may not show such specificity.

Bromonitro Compounds

Brominated compounds such as 2-bromo-2-nitropropane-1,3-diol and 5-bromo-5-nitro-1,3-dioxane (Bronopol and Bronidox, Boots Co. PLC) are effective against both bacteria and fungi at low concentrations. Antimicrobial activity resides in the $Br\text{-}C\text{-}NO_2$ group.[29]

Shepard et al.[30] examined the effects of Bronopol on both intact bacterial cells and isolated enzymes. Bronopol inhibited the oxidation of various substrates by intact cells. Bronopol inhibited glucose, acetate, and pyruvate oxidation by 50% at a concentration of 15 μg/ml. Higher Bronopol concentrations inhibited alpha-ketoglutarate oxidation by 50%.

In vitro effects of Bronopol on isolated dehydrogenase enzyme systems varied considerably. Noncompetitive, competitive, and uncompetitive inhibition of different enzymes systems was seen. The authors suggested that oxidation of thiol groups might account for the uncompetitive inhibition. Subsequent work confirmed the reactivity of Bronopol with thiol groups.[31] In the presence of Bronopol, cysteine converted to cystine. Bronopol appeared to play a catalytic role in this system. The concentration of Bronopol did not decrease during the conversion of cysteine to cystine.

Ghannoum et al. examined the antimicrobial mode of action of Bronidox in detail.[32] Bronidox converted enzyme thiol groups to disulfides, resulting in inactivation of enzymes with sulfhydryl groups at the active sites. Exposure of intact bacteria to Bronidox led to a decrease in respiratory activity, as seen by decreased O_2 uptake. Bronidox also inhibited nucleic acid synthesis as seen by a decreased incorporation of ^{3}H-uridine.

Ghannoum presented a sequence of 3 steps involved in the reaction of Bronidox with thiol groups: (1) The thiol group extracts bromine from Bronidox to form a thiobromo derivative. (2) The thiobromo derivative reacts with another thiol group to yield a disulfide and free bromide. (3) After the loss of bromine from the Bronidox molecule, gain of a proton yields the bromo derivative. The 2:1 stoichiometry observed in the *in vitro* reaction of thiol compounds with Bronidox is consistent with this mechanism.

In summary, available data indicate that bromonitro compounds react with essential cell enzymes. Therefore, these materials fall into the second mechanism category of Davidson and Branen.[8]

Chelating Agents

Ethylenediamine tetracetic acid (EDTA) and other chelating agents are weak antimicrobial agents by themselves. However, they can potentiate the activity of many other antimicrobial agents. Therefore, their mode of action deserves consideration in a discussion of the modes of action of antimicrobial agents.

Several chelating agents have been investigated for their preservative properties in cosmetic products.[33] Only EDTA and various potassium and sodium salts of this compound are the chelating agents in widespread use in cosmetic products.[1] In addition to exerting an antimicrobial benefit, EDTA has other properties that benefit cosmetic products.[34]

EDTA is most effective in improving the efficacy of other preservatives against Gram-negative bacteria, the more frequent agents of spoilage of liquid products. The antimicrobial properties of EDTA arise from the ability of this compound to chelate metals. Chelation of magnesium is the most significant for EDTAs antimicrobial effects.[35]

The outer membranes of Gram-negative bacteria contain lipopolysaccharide polymers that have a net negative charge. Positively charged magnesium ions associated with these polymers provide a cross-linking function and neutralize the repulsion of the polymers with like charges. Thus, these divalent cations contribute to the permeability barrier of Gram-negative bacteria. This permeability barrier provided by the outer membrane probably accounts for the substantial resistance to antimicrobial agents demonstrated by many Gram-negative bacteria, particularly *Pseudomonas* strains.

EDTA has a strong affinity for divalent ions and can remove them from cell membranes. Loss of these ions causes the cell membrane to increase in permeability, allowing other antimicrobial agents to penetrate the interior of the cell where they can exert their lethal effect. This membrane disruptive effect produced by EDTA exposure alone is sufficient to kill some microorganisms.[36] The more important effect of EDTA lies in making the exposed cells more sensitive to other antimicrobial agents.

SUMMARY

The mechanism of action of antibiotics is usually known on a very detailed level. This understanding is frequently a requirement for approval for human use of the drug. In contrast, far less is known about the action of many cosmetic preservatives. This area has received very little research except the laboratory of P. Gilbert.

Many antimicrobial agents used as cosmetic preservatives are bulk chemical commodities. Therefore, there is little impetus for the manufacturers to investigate the mechanisms of action of these compounds. This area has been outside the mainstream of biomedical research. There has been little interest in these compounds in the academic community.

Finally, the mode of action of cosmetic preservatives may be less specific than that of antibiotics. Preservatives damage many components of the microbial cell. It is difficult to pinpoint the specific event that is responsible for microbiocidal or microbiostatic activity of preservatives. In contrast, antibiotics usually derive their efficacy from interactions with a single enzyme or macromolecule.[37]

Davidson and Branen[8] established three general categories to cover the mechanisms of all antimicrobial agents. The first category includes compounds that cause a loss of membrane permeability. Loss of membrane permeability results in leakage of cell contents and a loss of energy production from the electron motive force.[3] The second category includes those agents that inhibit or inactivate essential cell enzymes. These compounds frequently attack the enzymes by reacting with thiol or hydroxyl groups at enzyme active sites. The third category involves the destruction of the genetic material. These compounds prevent replication of the cell. If sufficiently extensive, they disrupt synthetic and respiratory activities. A review by Hugo[38] provides further perspective on the development of these concepts.

Antimicrobial compounds destroy or inhibit microorganisms as a result of interactions with target sites on the microbial cell. The target site can be a specific cell structure or macromolecule affected by the antimicrobial compound. The damage suffered by the structure or macromolecule results in inhibition or death of the cell. The target site may be very specific or nonspecific. With antibiotics, the target site is frequently one specific protein.

The specificity of the target apparently declines as the reactivity of the antimicrobial increases. For example, strongly reactive compounds such as hypochlorite or hydrogen peroxide can react at many different sites in the microbial cell. The death of the cell is probably the summation of the damage inflicted at these various sites. Some antimicrobials such as the isothiazolinones may be more selective in their reactivity.

Exposure of a microbial cell to an antimicrobial agent causes several physiological changes in the cell. Some of these changes are primary effects where the antimicrobial agent directly interacts with the target site. Other physiological changes may be an indirect result of the damage caused by the antimicrobial agent. For example, respiration rates may decrease after exposure to the antimicrobial agent. Decreased respiration may not be due to a direct effect of the antimicrobial agent on respiratory enzymes. Instead, decreased respiration may be due to loss of permeability control derived from membrane damage caused by the antimicrobial agent. Determining the primary effects of an antimicrobial agent is often crucial to choosing the correct biocide or biocides to preserve a product.

REFERENCES

1. Decker, R. L. and Wenninger, J. A. (1987). Frequency of preservative use in cosmetic formulas as disclosed to FDA-1987, *Cosmet. Toilet.* 102, 21.
2. Denyer, S. P., Hugo, W. B., and Harding, V. D. (1985). Synergy in preservative combinations. *Int. J. Pharm.* 25, 245.
3. Furr, J. R. and Russell, A. D. (1972). Some factors influencing the activity of esters of p-hydroxybenzoic acid against *Serratia marcescens, Microbios* 5, 189.
4. McQuillen, K. (1960). Bacterial protoplasts, *The Bacteria,* vol. 1 (I. C. Gunsalus, R. Y. Stanier, Eds.), Academic Press, New York.
5. Freese, E., Chingju, W. S., and Galliers, E. (1973). Function of lipophilic acids as antimicrobial food additives. *Nature* 241, 321.
6. Ramos, S., Schuldiner, S., and Kaback, H. R. (1976). The electrochemical gradient of protons and its relationship to active transport in *Escherichia coli* membrane vesicles, *Proc. Natl. Acad. Sci. U.S.A.* 73, 1892.
7. Eklund, T. (1985). The effect of sorbic acid and esters of p-hydroxybenzoic acid on the protonmotive force in *Escherichia coli* membrane vesicles, *J. Gen. Microbiol.* 131, 73.
8. Davidson, P. M. and Branen, A. L. (1981). Antimicrobial activity of non-halogenated phenolic compounds, *J. Food Protection* 44, 623.
9. Suter, G. M. (1941) *Chem. Rev.* 28, 269.
10. Prindle, P. F. (1983). Phenolic compounds, *Disinfection, Sterilization, and Preservation,* third ed. (S. S. Block, Ed.), Lea and Febiger, Philadelphia, p. 197.

11. Hugo, W. B. and Bowen, J. G. (1973). *Microbios* 8, 189.
12. Hugo, W. B. (1978). *Int. J. Pharm.* 1, 127.
13. Silver, S. and Wendt, L. (1967). Mechanism of action of phenethyl alcohol: breakdown of the cellular permeability barrier.˙*J. Bacteriol.* 93, 560.
14. Treick, R. W. and Konetzka, W. A. (1964). Physiological state of *Escherichia coli* and the inhibition of deoxyribonucleic acid synthesis by phenethyl alcohol. *J. Bacteriol.* 88, 1580.
15. Hall, A. L. (1984). Cosmetically acceptable phenoxyethanol, *Cosmetic and Drug Preservation, Principles, and Practice* (J. J. Kabara, Ed.), Marcel Dekker, Inc., New York.
16. Gilbert, P., Beveridge, E. G., and Crone, P. B. (1976). *J. Pharm. Pharmacol. Suppl.* 28, 51.
17. Gilbert, P., Beveridge, E. G., and Crone, P. B. (1977). *Microbios* 19, 17.
18. Warth, A. D. (1977). Mechanism of resistance of *Saccharomyces bailii* to benzoic, sorbic, and other weak acids used as food preservatives, *J. Appl. Bacteriol.* 43, 215.
19. Warth, A. D. (1988). Effect of benzoic acid on growth yield of yeasts differing in their resistance to preservatives, *Appl. Environ. Microbiol.* 54, 2091.
20. Law, A. B., Moss, J. N., and Lashen, E. S. (1984). Kathon CG, a new single-component, broad spectrum preservative system for cosmetics and toiletries, *Cosmetic and Drug Preservation, Principles and Practice* (J. J. Kabara, Ed.) Marcel Dekker, Inc., New York, p. 129.
21. Fuller, S. J., et al. (1985). The mode of action of 1,2-benzisothiazolin-3-one on *Staphylococcus aureus, Letters in Applied Microbiology,* 1, 13.
22. Collier, P. J., Austin, P., and Gilbert, P. (1990). Uptake and distribution of some isothiazolone biocides into *Escherichia coli* ATCC 8739 and *Schizosaccharomyces pombe* NCYC 1354, *Int. J. Pharm.* 66, 201.
23. Collier, P. J., Ramsey, A., Waigh, R. D., Douglas, K. T., Austin, P., and Gilbert, P. (1990). Chemical reactivity of some isothiazolone biocides, *J. Appl. Bacteriol.* 69, 578.
24. Collier, P. J., Ramsey, A., Austin, P., and Gilbert, P. (1990). Growth inhibitory and biocidal activity of some isothiazolone biocides, *J. Appl. Bacteriol.* 69, 569.
25. Giles, C. H., Smith, D., and Huitson, A. (1974). A general treatment and classification of the solute adsorption isotherm. Part I. Theoretical, *J. Colloid Interface Sci.* 47, 755.
26. Weaver, J. E., Cardin, C. W., and Maibach, H. I. (1985). Dose-response assessments of Kathon biocide. *Contact Dermatitis* 12, 141.
27. Rosen, M. and McFarlan, A. G. (1984). Free formaldehyde in anionic shampoos, *J. Soc. Cosmet. Chem.* 35, 157.
28. Habeeb. (1968). *Arch. Biochem.* 126, 16.
29. Lappas, L. C., Hirsch, C. A., and Winely, C. L. (1976). Substituted 5-Nitro-1,3-dioxanes. Correlation of chemical structure and antimicrobial activity, *J. Pharmaceut. Sci.* 65, 1301.
30. Shepard, J. A., Woodcock, P. M., and Gilbert, P. (1985). Interaction of 2-bromo-2-nitropropan-1,3-diol with some respiration and dehydrogenase enzyme systems in *Escherichia coli, J. Pharm. Pharmacol.* 37(Suppl), 95P.

31. Shepherd, J. A., Buerby, M. R., Pemberton, D., Gilbert, P., and Waigh, R. D. (1987). The reactions of Bronopol (2-bromo-2-nitropropan-1,3-diol) with thiols, *J. Pharm. Pharmacol.* 39(Suppl), 116P.
32. Ghannoum, M., Thompson, M., Bowman, W., and Al-Khalil, S. (1986). Mode of action of the antimicrobial compound 5-Bromo-5-nitro-1,3-dioxane (Bronidox), *Folia Microbiol.* 31, 19.
33. Hart, J. R. (1984). Chelating agents as preservative potentiators, *Cosmetic and Drug Preservation, Principles and Practice* (J. J. Kabara, Ed.), Marcel Dekker, Inc., New York, p. 129.
34. Hart, J. R. (1983). EDTA-type chelating agents in personal care products, *Cosmetics and Toiletries* 98, 54.
35. Wilkinson, S. G. (1975). Sensitivity to ethylenediaminetetracetic acid, Resistance of *Pseudomonas aeruginosa* (M. R. W. Brown, Ed.), John Wiley & Sons, New York, p. 145.
36. Eagon, R. G. and Carson, K. J. (1965). Lysis of cell walls and intact cells of *Pseudomonas aeruginosa* by ethylenediamine tetraacetic acid and by lysozyme, *Can. J. Microbiol.* 11, 193.
37. Lambert, P. A. (1987). Mechanisms of action of antibiotics, *Pharmaceutical Microbiology*, 4th ed. (W. B. Hugo, A. D. Russell, Eds.), Blackwell Scientific, Oxford, p.186.
38. Hugo, W. B. (1976). Survival of microbes exposed to chemical stress, *Symposium of the Soc. for General Microbiol.* 26, 383.

SECTION FOUR: REGULATORY AND TOXICOLOGY CONCERNS

8 CONSUMER SAFETY CONSIDERATIONS ASSOCIATED WITH THE MICROBIAL PRESERVATION OF COSMETICS

William A. Apel

CONTENTS

INTRODUCTION

The microbiologist developing preservative systems for cosmetic products confronts several technical challenges. These include identifying a preservative that is (1) efficacious, (2) stable for the life of the product, (3) cost effective, (4) compatible with the product matrix, (5) acceptable from a regulatory view, and (6) safe for the consumer. Meeting these criteria represents a formidable task. Consumer safety considerations ultimately

0-8493-3713-5/97/$0.00+$.50

determine whether or not one uses an otherwise acceptable preservative in a marketed product.

What do we mean by "safety considerations"? Safety considerations pertain to whether or not the product will be safe for the consumer under normal use and foreseeable misuse conditions. For the microbiologist, this means that he or she must protect the product from significant microbial contamination. This must be done with preservatives that are toxicologically safe in the product when used by consumers.

Other portions of this book deal extensively with preservative efficacy issues. This chapter will provide the cosmetic microbiologist with a background in toxicology for the development of cosmetic preservatives. This information will not, however, prepare the microbiologist to develop his or her own independent preservative safety testing program. Instead, it will provide a basic understanding that will be useful when interacting with toxicologists responsible for preservative safety evaluations.

The information contained in this chapter will also heighten the cosmetic microbiologist's awareness of preservative safety as a major issue. This awareness will help decrease the time lost and expense associated with disqualifying good preservatives due to safety concerns.

The safety perspective has become especially important in recent years. Consumers have become more sophisticated in their knowledge of cosmetic ingredients, including preservatives. They are also far more aware of the test methods used to assure ingredient safety. This knowledge base ranges across a variety of issues. Two of the key ones are (1) "hit lists" of various ingredients judged by special interest groups to be potentially hazardous, and (2) a growing desire to lessen or eliminate animal testing associated with the toxicology testing of cosmetic products and their ingredients.

Through corporate public relations, these issues impact directly on the development of cosmetic preservative systems. As a result, the microbiologist responsible for preserving cosmetic products in this complex scientific and social environment should develop a basic understanding of relevant toxicological considerations relative to preservatives.

USE OF EXISTING INFORMATION

When faced with the task of developing a preservative system for a new product, the microbiologist typically first considers using preservatives that are commercially marketed for cosmetic products. During this process, the preservative manufacturer is often contacted in an effort to obtain information concerning the efficacy of the preservative in question. In the course of obtaining efficacy information, the microbiologist should also specifically request any information that the manufacturer has available concerning toxicological testing the manufacturer has already performed with the preservative. Often, considerable toxicological information is immediately available at no cost from the manufacturer. While the types of safety test information vary widely from manufacturer to manufacturer, at a minimum they should have results from primary skin irritation, sensitization, and basic mutagenicity testing. More extensive safety test data will often be obtainable from the manufacturer of a well established, high sales volume preservative.

When secured early in the development cycle, a qualified toxicologist can use these data to help determine whether significant safety issues appear to exist with the preservative in question. The microbiologist can save considerable time and expense by not performing extensive microbiological evaluations with a preservative that might not be toxicologically suitable for the product application under consideration. Likewise, if the safety profile provided by the manufacturer indicates no apparent toxicological problems, the microbiologist can commence efficacy testing with a higher degree of confidence that, upon successful completion of this testing, unforeseen toxicological issues will not preclude the use of the preservative.

If there are insufficient safety test data, it may be necessary to begin microbial efficacy testing of a preservative while safety testing is concurrently being conducted. An explanation of typical safety tests conducted with cosmetic preservatives, together with a fundamental background in the principles of toxicology, is given below.

FUNDAMENTALS OF TOXICOLOGY

Toxicology is the study of deleterious effects of chemical, physical, or biological agents on living organisms. The degree to which these deleterious effects are elicited is dependent on a number of factors including the species/strain of the organism exposed, the stage of development of the organism, the compound/physical effect to which the organism is exposed, the duration of exposure, frequency of exposure, route of exposure, site of exposure, etc. A brief discussion of these factors is as follows.

The toxicity of a compound such as a preservative can vary significantly from species to species. There are many reasons for this variance, but for the most part they are a product of the organism's physiology and anatomy, both of which tend to control the organism's actual exposure to the toxicant. For instance, an animal with a relatively thin, fragile skin is more likely to be subject to the effects of a topically applied toxicant than an animal with a thicker, less penetrable skin. Likewise, if an otherwise emetic toxicant is orally administered, an animal incapable of complete regurgitation such as a rat is more likely to display toxic effects than is an animal like a dog which can more completely regurgitate. Similarly, an organism capable of rapid and complete metabolism or excretion of a toxicant is much less prone to exhibiting toxic effects than is an organism incapable of effectively eliminating the toxicant from its system. Thus, for these reasons, it is important to judiciously select the species with which toxicologic testing will be performed. Normally, when considering cosmetic toxicology, this selection either attempts to mimic or amplify the effect expected in humans, depending on the goal of the test.

Toxicity of chemical and physical agents can vary over a wide range. For example, one of the most toxic compounds known to man, botulism toxin, which is produced by the bacterium *Clostridium botulinum*, has an LD_{50} (see explanation for LD_{50} under acute toxicity testing below) of 10 ng kg^{-1} and is classified as supertoxic. Sodium chloride, on the other hand, has an LD_{50} of approximately 4 g kg^{-1} and is classified as only moderately toxic.

Route of exposure can also drastically influence the toxicity exhibited by a compound. Major routes of exposure are topical (a particularly important exposure route in the case of cosmetiçs), inhalation, ingestion, and parenteral. Typically, the maximum toxic effect is observed by whichever route (based on the chemistry of the toxicant and on the physiology and anatomy of the organism) will gain the toxicant the quickest and most prolonged presence in the circulatory system.

In a similar manner, site of exposure can also alter the toxic effect of a compound. For example, in humans a compound would be more likely to transit the skin-blood barrier at a site where the skin is thin such as the post mandibular area than it would at a site where the skin is thick and less subject to absorption such as the soles of the feet. It should be noted that the matrix in which the toxicant is applied can significantly effect toxicity by affecting absorption of the toxicant by the organism. Specifically, the chemical composition of the matrix, the amount of matrix, and the concentration of the toxicant in the matrix also serve to influence percutaneous absorption. This is an especially important consideration when testing preservatives in cosmetic products which are composed of complex matrices as these preservatives may feasibly enter the body via percutaneous absorption.

Duration of exposure also can influence toxic effects. Toxicologists group their exposure testing into four categories (Table 8-1).

TABLE 8-1

DURATION OF EXPOSURE

Exposure Category	**Duration**
Acute	<24 hrs
Subacute	1 to 30 days
Subchronic	30 to 90 days
Chronic	>90 days

Frequency, as well as level of exposure, can significantly alter toxic response to a compound. For example, an acute, high concentration oral exposure to chromium can lead to severe renal necrosis.[1] A much lower, but more frequent, exposure to chromate can lead to carcinogenic effects[2] and skin sensitization.[3] Similarly, a single acute dose of a compound may lead to a single toxic effect. However, even chronic doses at low levels with the same compound may not be toxic if the organism has enough time to metabolize or excrete it before it reaches effect levels. If the compound accumulates in tissues, lower chronic doses may be additive and exhibit the same effect as a single high level acute exposure. As a result, duration, frequency, and level of exposure are all important factors to consider in planning a cosmetic preservative testing program for toxicology.

GENERAL TOXICITY TESTS

There are several different tests used to assess the toxicity of preservative compounds either as neat compounds or in product. In the low cost and short term tests, toxicologists can evaluate preservatives in both forms. There is a tendency to conduct testing in long term, more expensive subchronic and chronic tests with preservatives at various levels in product. There is no definitive rule, however, whether one should test a preservative in pure form or in a product matrix. A qualified toxicologist needs to determine this on a case by case basis.

In this section, several different general toxicology tests were summarized. More detail of the tests most relevant to preservative development are provided separately in the following sections of this chapter.

Acute Toxicity Tests

The acute toxicity test is the first and among the most economical of the tests performed on any preservative compound. As such, data from these tests are often available from the preservative manufacturer. Toxicologists conduct acute toxicity tests such as LD_{50} using rats or mice. In the instance of dermal exposure, toxicologists use rabbits. The amount of a compound that will kill 50% of the population is the *L*ethal *D*ose 50% or LD_{50}.

Route of exposure can have an important effect on the apparent toxicity of the compound. Probably the most universal route of exposure is oral, with additional testing done using the exposure route anticipated with the product in question. With cosmetics, this is often dermal. In acute tests, toxicologists expose animal populations to single doses of various amounts of the test compound. The animals are then observed for signs of toxicity for a defined period, often 14 days. The toxicologist reports the LD_{50} as the mass of compound (i.e., preservative) per mass of animal body weight. Table 8-2 summarizes typical toxicity ratings.

TABLE 8-2

TOXICITY RATINGS

Toxicity rating	Dosage
Super toxic	<5 mg kg^{-1}
Extremely toxic	5 to 50 mg kg^{-1}
Very toxic	50 to 500 mg kg^{-1}
Moderately toxic	0.5 to 5 g kg^{-1}
Slightly toxic	5 to 15 g kg^{-1}
Essentially nontoxic	>15 g kg^{-1}

In addition to establishing a comparative LD_{50} value, the LD_{50} information is useful in determining preliminary dosing regimens for more extensive toxicity tests. It also helps clarify the clinical manifestations of toxicity for the compound in question.

Cosmetic manufacturers have made significant strides in lessening the number of animals sacrificed in acute toxicity testing. They have done this by developing more statistically founded protocols to determine LD_{50} with less animals. The development and use of *in vitro* toxicity testing protocols may help eliminate the use of live animals entirely.

Subacute Toxicity Tests

Subacute toxicity tests are similar to acute toxicity tests except the toxicologist doses the animal populations repeatedly over a short interval (e.g., 2 weeks) with the test compound. The animals are observed for signs of toxicity. This observation may include gross observations, histopathological observations, and chemical evaluations. The information obtained from this type of testing is useful in establishing dosage protocols for longer term toxicity testing.

Subchronic Toxicity Tests

Subchronic toxicity testing is longer duration testing; a typical test lasts about 90 days. The test compound is given via the route of anticipated consumer exposure. With cosmetics, this is often dermal with the test animals being rabbits. If little is known about the subchronic toxicity of the compound or product, then one also does an orally administered test in rodents. Toxicologists use several different dosage levels in subchronic testing. They base the dosage levels on results from subacute testing. Typically, one selects the dosage levels so the highest dose will cause a low level of mortality while the lowest dose will cause no grossly observable toxic effects. Observations made during and immediately following subchronic testing include mortality, body weights, food consumption rates, various chemical parameters, hematology, and histology.

Chronic Toxicity Tests

As the name implies, chronic toxicity testing is subchronic testing conducted over a longer interval. Test duration may vary from 180 days up to 2 years or more to mimic a lifetime exposure in humans. Chronic testing is especially useful in detecting carcinogens.

Ocular Irritation Tests

Ocular irritation tests are often conducted on cosmetic products. The best known of these tests is the Draize test. There are many variations on the Draize test.[4] All of them involve instillation of the test compound into a rabbit's eye and then grading the eye for irritancy. This type of testing has come under severe criticism by animal rights groups. This criticism has led to modification of the Draize protocol to use lower doses of compounds to reduce the resulting irritation. In recent years, considerable experimentation with *in vitro* eye irritation tests has also been done.

SKIN TOXICITY TESTING

Background

The skin is an organ that serves as a primary interface between the body and the outside environment. As such, it provides a barrier to external physical and chemical agents with a primary function of maintaining the composition and constituency of the body. Physically, the skin is composed of two layers, the outermost being the epidermis and the innermost being the dermis.

The epidermis has four discrete layers containing several distinct cell types including Langerhans cells (epidermal macrophages important in processing allergens), keratinocytes, and melanocytes. The outer epidermal layer, or stratum corneum, consists of flattened, dried keratinocytes. The cells of the stratum corneum are metabolically inactive and derived from the deeper, metabolically active layers of the epidermis. The stratum corneum resists mechanical stretching and acts as a primary barrier in the skin against the transport of chemical substances. Once past the stratum corneum, there is a minimal barrier protecting against transporting chemicals into the deeper layers of the skin.

Below the stratum corneum lie three layers of metabolically active cells: the stratum granulosum, the stratum spinosum, and the stratum germinativum. The stratum germinativum is attached to the dermis by the basal lamina. The basal lamina attaches the epidermis to the dermis. It also has several epidermal appendages composed of modified epidermal cells extending from the epidermis into the dermis. Hair follicles, apocrine and eccrine sweat glands, and sebaceous glands are all examples of epidermal appendages.

The dermis consists of several connective tissues including collagen, elastin, and reticulin. This combination gives the dermis a loose, elastic quality. The dermis has many blood vessels, and thus if traumatized, can produce external bleeding. As a result of its vascularization, immunologically active cells including macrophages, mast cells, and lymphocytes are found in the dermis.

Percutaneous absorption of chemical compounds (including preservative agents) by the skin is a primary factor contributing to both local and systemic toxic effects. As a result, the absorptive characteristics of a chemical are often a significant factor in whether or not that chemical will produce dermal toxicity. Many nonpolar, lipophilic compounds readily diffuse across the skin barrier.[5]

Diffusion of hydrophilic, polar compounds is highly influenced by the hydration state of the stratum corneum. Depending on environmental conditions, the stratum corneum can be hydrated with 10 to 70% water. As the stratum corneum becomes hydrated, hydrophilic compounds diffuse easily across the barrier. Therefore, the conditions under which one does skin testing (e.g., closed patch, open patch, etc.) can significantly affect the outcome the test.

The epidermal appendages also provide a mechanism for diffusion of chemicals across the skin. This effect is probably secondary to transepidermal diffusion, however, since these appendages make up only a small fraction of the skin's total surface area.

The work of Feldmann and Maibach[6] have shown that percutaneous absorptive characteristics are highly dependent on skin regions. In studies with radiolabeled hydrocortisone, they showed that the skin of the plantar foot arch is 300 times less sorptive than scrotal skin. By comparison, if the relative sorption of the foot arch skin was set equal to 1, the skin of the forearm was 7 to 8 times more absorptive. The skin of the back was 12 times more absorptive. The arm and back absorption rates are of particular interest since they are the sites most frequently chosen for skin toxicity tests.

The vehicle chosen for testing a preservative can have significant effects on the test's outcome. This is because the vehicle may enhance percutaneous absorption. The choice of highly lipophilic, nonpolar vehicles can drastically increase skin penetration. Surfactant containing vehicles, particularly those with either anionic or cationic surfactants (like those found in shampoos and soaps), can drastically increase percutaneous absorption. The mechanism of action for these agents is associated with chemical disruption of the stratum corneum, which leads to increased permeability.

Toxic Skin Responses to Preservatives

In general, one can expect two categories of toxic skin response with preservative agents. These are nonimmunologically mediated irritant responses and immunologically mediated allergic responses. Since the mechanisms for these responses are different, they will be discussed separately.

Irritant Responses

These are elicited with antimicrobial preservatives in two broad categories: (1) corrosion and (2) irritation. Corrosion is irritation resulting from the direct necrotic action of the chemical on skin with an irreversible disintegration of the skin tissue. It often results in scarring. Chemical burns, such as those received from strong acid or bases, would be classic examples of corrosion.

Irritation is divided into three categories: (1) acute irritation, (2) cumulative irritation, and (3) photoirritation. Acute irritation occurs as a localized, reversible noninflammatory response resulting from a single application of a toxicant. Cumulative irritation results from repeated exposure to toxicants that, upon initial application, do not elicit acute irritation. Phototoxicity results from light-induced chemical changes in a compound once applied to the skin such that the compound becomes an irritant.

In practice, many antimicrobial preservatives that elicit an irritation response act as either acute or cumulative irritants. Typically, one tests for this type of irritancy using either animal or human patch tests. In these tests, one places varying concentrations of the preservative under occlusive patches on the skin of either animals like rabbits or on the arms or backs of human volunteers.

For acute irritancy testing, toxicologists often use a prophetic patch testing protocol.[7] In this type of testing, one places the material in an occlusive patch onto the test subject for 24 to 48 hours. At the end of the test period, the patch is removed and the skin allowed to equilibrate for 30 to 60 minutes. The skin is then scored for amounts of erythema and edema. One can use the International Contact Dermatitis Research Group Scoring Scale[8] to score the results. Table 8-3 summarizes this four point scale.

Toxicologists use the cumulative irritancy patch test to test chemicals for additive irritancy.[9] This test is a variation on the prophetic patch test outlined above except one conducts the test from 14 to 21 days with

TABLE 8-3

INTERNATIONAL CONTACT DERMATITIS RESEARCH GROUP SCORING SCALE FOR IRRITATION TESTING

Description	Score
No reaction	0
Erythema	1+
Erythema and edema	2+
Marked erythema and edema	3+

repeated applications of the test material. The interval used in this test is a point of contention. Originally, the test was a 21 day test. In recent years, however, the test has been shortened to 14 days. This decreases costs and reduces chronic skin fatigue that may occur in a traditional 21 day test. Skin fatigue is undesirable since it makes accurate interpretation of results extremely difficult. In addition, research has shown that approximately 95% of all reactions take place in the first 14 days of the 21 day period.[10] These factors taken together constitute a strong argument for the 14 day test format.

Skin Sensitization

Also known as allergic contact dermatitis, this is a type IV response according to the scheme of Coombs and Gell.[11] The key factors characterizing type IV reactions are that they are delayed and are T lymphocyte and macrophage mediated reactions. This distinguishes them from the other three categories of allergic responses (i.e., types I to III) in that these other reactions are primarily antibody mediated. The allergic contact dermatitis that results from exposures to poison ivy is a classic example of the delayed type IV hypersensitivity (DTH).

From a simplistic standpoint, a type IV skin reaction proceeds as follows. The skin absorbs an allergen (typically a low molecular weight hapten compound) which conjugates with the skin proteins. Langerhans cells in the epidermis then react with the allergen-protein complex, process it, and present it to afferent, precursor T lymphocytes. The T lymphocytes then become "educated" effector T lymphocytes sensitized to the specific allergen. This first portion of the type IV skin reaction is induction.

The second stage is elicitation, when sensitized T lymphocytes are exposed to the allergen and produce lymphokines. Lymphokines are soluble, chemical factors that include macrophage activation factor (MAF), interferon, skin reactivity factor (SRF), and chemotactic factor (CF). The reactions mediated by these lymphokines result in localized symptoms such as erythema and edema that are characteristic of type IV responses.

In addition to catechols from poison ivy, many other compounds are known to induce skin sensitization. Among these compounds are several cosmetic preservatives that can induce hypersensitivity under test conditions and actual use conditions. Table 8-4 shows these compounds.[12,13]

Toxicologists usually test suspected sensitizers for their sensitization potential using a patch test that parallels the one described for irritants. One performs this patch testing in either animals (mostly guinea pigs, a species that possesses a satisfactory type IV skin response) or humans. One typically uses animals for preliminary tests to gain a gross judgment on sensitization potential. If the compound proves acceptable in animal testing, one can then do human testing on volunteers to gain a more definitive idea of sensitization potential.

TABLE 8-4

EXAMPLES OF POTENTIAL CONTACT SENSITIZERS USED FOR PRESERVATIVE APPLICATIONS IN COSMETICS

Formaldehyde
Paraben esters
Sorbic acid
Isothiazolines
Organic mercurial compounds
Phenolics (e.g., hexachlorophene)

Some potential sensitizers may also be irritants. If so, one can use lower concentrations of the compound to hold irritation to a minimum. In practice, this is done by conducting cumulative irritation testing before starting sensitization tests. This allows the toxicologist to do the sensitization testing at levels lower than those known to cause irritation.

There are also two basic stages to the patch test protocol for sensitization. The first stage is the induction period where single or multiple patches, which contain the test agent, are applied to the subject. The second stage is the elicitation stage. It is done after a brief rest period of about two weeks following the final induction patching. It consists of reapplying a challenge patch containing the test agent to the subject. After removal of the challenge patch, the toxicologist grades the subject at several times (24, 48, 72, and 96 hrs) for symptoms of a positive or negative sensitization response. These symptoms consist of erythema, edema, itching, and in some cases blistering.

Concentrations of potential allergen applied can be of extreme importance in the above test format. For example, one can use a higher level of test compound to induce sensitization than that required in a presensitized population. Thus, common sense must enter sensitization test design. Testing a commonly used preservative for sensitization potential in a given product is a good example.

The toxicologist should be aware that the consumer population may already have been exposed to the preservative in a variety of applications. Some of the exposures may even have been at high levels. In this case, one must determine the induction level of the preservative as well as the elicitation level. One can determine this latter level by challenge testing of a presensitized population. If the assumption is made that a sensitized population exists among the consumer population exposed to the preservative, a different result will be obtained than if we do not make such an assumption. As such, elicitation testing is a primary concern and the term *index of sensitivity* is used to describe the prevalence of sensitivity in the population.[14]

If the preservative under consideration is a proprietary agent to which the consumer population has had no previous exposure, only sensitization induction levels are of concern. Without exposure to the compound or close analogues, we do not expect that the consumer population is sensitized.

Use Testing for Sensitization

This is a technique used in evaluating products containing preservatives showing borderline sensitization potential during patch testing. This technique can be particularly useful in evaluation of the elicitation potential of a preservative when using a presensitized test population. In this type of testing, an informed volunteer is given production samples and instructed to use the product as they typically would for personal use. At the beginning of the test, panel members are told to report any erythema, edema, or unusual itching that occurs. In addition, panelists are periodically inspected during the test for signs of sensitization. If the product containing the preservative elicits a response, it is likely that a substitute preservative with less sensitization potential will be needed.

If no response is seen under actual test conditions with a large test population, then one has supporting evidence that the preservative as used in the product may be safe for marketing. The results from this use testing help decide whether a preservative that shows marginal sensitization potential during patch testing is acceptable for use in marketed cosmetics.

GENETIC AND DEVELOPMENTAL TESTING

Mutagenicity Testing

The realistic assessment of mutagenic potential for preservative compounds is of special relevance to cosmetic preservatives. Many preservative compounds have tested positive for mutagenicity in certain mutagenicity tests, particularly those conducted *in vitro*. Simplistically, mutagenesis is the process by which a chemical or physical agent causes an alteration in the germ-line cells. These are the cells responsible for genetic transfer from generation to generation such as sperm and eggs. Mutations can occur in either somatic or germ cells. In germ cells, malformation or death of the resulting embryo can occur. In somatic mutations, cancer may be the result. Thus, mutagenic testing is important to establish the mutagenic potential of a preservative as well as to screen for potential carcinogens. This latter concern is based on the realization that initiation of carcinogenesis (i.e., transformation) is likely a mutagenic occurrence.

Mechanistically, mutagenesis involves changes in the DNA base pairs. These changes consist of addition or deletion of base pairs or base pair substitutions as illustrated in Figure 8-1. Addition or deletion of base pairs can alter the reading frame. This is a frameshift mutation.

Base Pair Substitutions
  Transversions
    GC to CG
    GC to TA
    AT to TA
    AT to CG
  Transitions
    AT to GC
    GC to AT

Base Pair Addition or Deletion
  Addition

–GGGGGG–	to	–GGGXXGGG–
–CCCCCC–		–GGGYYGGG–

Figure 8-1 Genetic mechanisms of mutagenesis.

Several different tests, *in vitro* and *in vivo*, are used to detect chemically induced mutations. These tests are indirect evaluations of the mutational event. They detect a phenotypic change due to the mutation vs. the actual alteration in the DNA itself.

Perhaps the most commonly conducted mutagenesis assay is the Ames test. It is also known as the *Salmonella* test.[15] This test employs Dr. Bruce Ames' operon mutant strains of *Salmonella typhimurium*. These revert by addition/deletion mutations or base pair substitutions to the wild-type strain not requiring any growth factors. The degree of correlation with the Ames test in detecting mutagens is nearly 83%.[16] As a result of this high degree of correlation coupled with its low cost, toxicologists often used the Ames test as a first screening test for chemical mutagens.

Toxicologists also use other *in vitro* screening tests with eukaryotic cell lines to evaluate chemical mutagenicity. These include tests with Chinese hamster ovary cells and human lymphoblast and fibroblast cell lines.[17,18] The basis of this test is similar to that of the Ames test: the mutation event detected is based on phenotypic change induced by addition/deletion or base pair substitution mutations.

In vivo mutagenicity tests are also commonly performed. Typical of these are the *Drosophila* and the mouse mutagenicity tests. The *Drosophila* tests assay for phenotypic changes such as alterations in eye color, wing morphology, or bristle appearance. The mouse mutagenicity tests assay for changes in coat color. These types of *in vivo* mutagenicity assays are more accurate predictors of the mutagenic potential of many chemical compounds. They are also less frequently run than are the *in vitro* tests due to the increased cost associated with them.

Developmental Toxicity Testing

Any agent that causes an adverse effect in a developing embryo or fetus is a developmental toxicant. Typical of developmental toxicants are teratogens, agents responsible for birth defects. The number of known teratogens is extensive. They include biological entities (rubella virus), physical entities (X-rays), and a diversity of chemicals including pharmaceuticals like thalidomide and diethylstilbestrol.[19–23] Many chemical agents cause teratogenic effects under sometimes very specific conditions. Biologically active molecules such as cosmetic preservatives, especially new types in developmental testing, are certainly subjects for teratology testing.

Manson et al.[24] has provided a thorough review of teratogenic testing procedures. For the evaluation of potential short term exposures, a single generation animal test using rodents or rabbits is often used. One usually conducts this testing in three parts.

In the first part, the animals are dosed before and during mating with continued dosing of the females during pregnancy and lactation. They are then sacrificed during pregnancy and the unborn offspring examined for abnormalities. Other animals are allowed to deliver. Delivered offspring are weaned and examined for abnormalities. In the second portion of testing, pregnant females are dosed only during organogenesis and sacrificed just before delivery. The unborn offspring are examined for abnormalities.

In the third part of the short term testing regimen, pregnant females are dosed during the final trimester of pregnancy and through the lactation period. Offspring may be examined for abnormalities immediately following weaning or be allowed to reach adulthood prior to examination.

The variation in dosing tries to compensate for the effects of exposure timing on the teratogenic potential of the test compound. It is well known that exposure across very specific and oftentimes narrow windows can greatly affect whether or not a compound is teratogenic. Furthermore, for a teratogenic compound the timing of the exposure may influence the type of teratogenic effect observed.

To evaluate the effects of longer term exposures to potential teratogenic agents, multiple generation studies proceeding through at least three generations are used. These studies are conducted much like single generation studies except that a portion of each generation is retained and mated to produce the next study generation. In this way, the effects of the test compound on congenital abnormalities as well as fertility and litter size, viability, and growth are determined.

RISK ASSESSMENT

The purpose of the tests described above is to have the data needed to perform a risk assessment on the preservative used in the product under intended use and foreseeable misuse conditions. What constitutes an acceptable risk, however, is highly subjective. One must establish this risk on a case by case basis. Several factors need to enter into making this judgement:

1. Benefits gained by using the preservative.
2. Chances of subjecting the consumer population to risks above those normally encountered in the course of daily living.
3. Ability to substitute alternative lower risk compounds for the same use.
4. Economic benefits derived.
5. Effects on the quality of life of the consumer.
6. Effects on the environment.
7. Effects on corporate public relations.

One can use established guidelines along with toxicologic test data to help in making risk assessments associated with the use of a given preservative.[25] One can use toxicology test data to establish two parameters: the lowest-observed-effect-level (LOEL) and the no-observed-effect-level (NOEL). The LOEL is the lowest dose at which a toxic effect was observed during testing. The NOEL is the highest dose at which no toxic effects were observed during testing.

The accuracy of these values depends on the validity of the test protocol used, how closely the doses tested were spaced, and the number of subjects tested. Usually after one establishes the NOEL values, safety factors are applied to these values. A safety factor of 10X is often used when one bases the NOEL data on valid chronic exposure data and large base size testing in humans (Klaasen, 1986). The 10X safety factor considers the variable responses found in human populations. One usually uses a safety factor of 100X when testing has been performed in animals. We include the extra 10X multiplier to account for interspecies differences. In instances where sufficient chronic exposure data are unavailable, a 1000X safety factor may be applied.

REFERENCES

1. Landgard, S. and Norseth, T. Chromium. (In) Friberg, L., Nordberg, G. F., and Vouk, V. B. (Eds). *Handbook on the Toxicology of Metals.* Elsevier, New York, NY, 1979.

2. Norseth, T. The carcinogenicity of chromium. *Environ. Health Perspect.* 40, 121, 1981.
3. Peltonen, L. and Fraki, J. Prevalence of dichromate sensitivity. *Contact Dermatol.* 9, 190, 1980.
4. Draize, J. H. and Kelley, E. A. Toxicity to eye mucosa of certain cosmetic preparations containing surfactant active agents. *Proc. Sci. Sect. Toilet Goods Assoc.* 17, 1, 1952.
5. Scheuplein, R. J. and Blank, I. H. Permeability of the skin. *Physiol. Rev.* 51, 702, 1971.
6. Feldmann, R. J. and Maibach, H. I. Absorption of some organic compounds through the skin in man. *Invest. Dermatol.* 54, 339, 1967.
7. Schwartz, L. and Peck, S. M., The patch test and contact dermatitis, *Public Health Report* 59:2, 1944.
8. International Contact Dermatitis Research Group. Terminology of Contact Dermatitis. *Acta Dermatol.* 50, 287, 1970.
9. Lanman, B. M., Zelvers, W. V., and Howard, C. S., The role of human patch testing the product development program. (In) *Proceedings of the Joint conference of Cosmetic Sciences, The Toilet Goods Association,* Washington, D.C., 1968.
10. Berger, R. S. and Bowman, J. P., Early appraisal of the 21 day cumulative irritation test in man. *J. Toxicol. Cut. Ocular Toxicol.,* 1, 109, 1982.
11. Combs, R. R. A. and Gell, P. G. H. Classification of allergic reactions responsible for clinical hypersensitivity and disease. (In) Gell, P. G. H., Coombs, R. R. A., and Lachman, P. J. (Eds). *Clinical Aspects of Immunology.* Oxford Press, Blackwell England, 1975.
12. Schorr, W. F. Cosmetic Allergy. A comprehensive study of the many groups of chemical antimicrobial agents. *Arch. Dermatol.,* 104, 459, 1971.
13. Bauer, R. L., Ramsey, D. L., and Bondi, E. The most common contact allergens. *Arch. Dermatol.* 108, 74, 1973.
14. Emmett, E. A. Toxic responses of the skin. (In) Klaassen, C. D., Amdur, M. O., and Doul J. (Eds). *Toxicology: The Basic Science of Poisons.* Macmillan Publishing Co., New York, NY, 1986.
15. Ames, B. N., McCann, J., and Yamasaki, E. Methods for detecting carcinogens and mutagens with the *Salmonella*/mammalian-microsome mutagenicity test. *Mutat. Res.* 31, 347, 1975.
16. Ames, B. N. and McCann, J. Validation of the *Salmonella* test: a reply to Rinkus and Legator. *Cancer Res.* 41, 4192, 1981.
17. Baker, R., Brunette, D., Mankovitz, R., Thompson, L., Whitmore, G., Siminovitch, L., and Till, J. Ouabain-resistant mutants of mouse and hamster cells in culture. *Cell* 1, 9, 1974.
18. Thilly, W. Analysis of chemically induced mutations in single cell populations. (In) Lawerance, W. (Ed): *Induced Mutagenesis: Molecular Mechanisms and their Implications for Environmental Protection.* Plenum Press, New York, NY, 1983.
19. Alford, C. A. Rubella. (In) Remington, J. S. and Klein, J. O. (Eds). *Infectious Diseases of the Fetus and Newborn Infant.* Saunders, Philadelphia, PA, 1976.
20. Gregg, N. M. Congential cataract following German measles in the mother. *Trans. Ophthalmol. Soc. NZ* 3, 35, 1941.

21. Warkany, J. Development of experimental mammalian teratology. (In) Wilson, J. G. and Warkany, J. (Eds.). *Teratology: Principles and Techniques.* University of Chicago Press, Chicago, IL, 1965.
22. McBride, W. G. Thalidomide and congenital anomalies. *Lancet* 2, 1358, 1961.
23. Postkanzer, D. and Herbst, A. Epidemiology of vaginal adenosis and adenocarcinoma associated with exposure to stilbestrol *in utero. Cancer* 39, 1892, 1977.
24. Manson, J. M., Zenick, H., and Costlow. R., Teratology test methods for laboratory animals. (In) Hayes, A. W. (Ed.) *Principles and Methods of Toxicology,* Raven Press, New York, NY, 1982.
25. Klaasen, C. D. Principles of Toxicology. (In) Klaassen, C. D., Amdur, M. O., and Doul, J. (Eds). *Toxicology: The Basic Science of Poisons.* Macmillan Publishing Co., New York, NY, 1986.

9 LAWS AND ENFORCEMENT

Mary K. Bruch

CONTENTS

COSMETIC REGULATIONS

Historical Perspectives

The Federal Food, Drug, and Cosmetic Act as we know it derives from the amendments to the Act passed in 1938 (1). The initial enactment of the Federal Food and Drug Act in 1906 was in response to the outrageous abuses in the food industry vividly portrayed in Upton Sinclair's *The Jungle.*

Although Congress was considering changes in the 1906 law, action leading to the 1938 Amendments occurred because of a serious accident. A manufacturer used ethylene glycol as a vehicle in elixir of sulfanilamide resulting in several deaths before sale of the product could be stopped.

0-8493-3713-5/97/$0.00+$.50

The furor raised by this tragedy resulted in quick empowerment with changes resulting in the 1938 Amendments.

A third important revision and additions to the Food, Drug, and Cosmetic Act came in 1968 in response to another disaster. Thalidomide was found to cause severe disfiguring abnormalities in unborn babies. These serious incidents added regulatory powers and improved reinforcement of the original act.

In contrast, though there have been some toxicities and seizures of cosmetic products, no major problems have stimulated increased regulation of cosmetics. We must realize that cosmetics are the least regulated of all the product categories under the jurisdiction of the FDA.

Because of the lighter hand in regulation, many believe that more serious regulation would provide better safety for the consumer. In contrast, many feel that the safety record of cosmetics has been so good that pre-clearance of cosmetics or more vigorous regulation is unnecessary. Occurrences over the last 50 years have balanced the see-saw to one side or the other. The debate continues.

At the heart of these divergent opinions is the risk-benefit balance that determines decisions concerning drugs and devices. The answers are easy when one is considering a cure for cancer, but not so easy when considering a "cure" from looking unattractive. The risks of a drug may be balanced when a life is at stake but unbalanced when one's perception of beauty is the consideration. When the benefit is minimal (smelling or looking better), no risk can be balanced in the equation. Each new incident propels the Food and Drug Administration to assess the risk-benefit ratio again. Often the benefit loses and the number of cosmetic ingredients shrinks.[2]

Over the last 50 years cosmetics have become a multi-billion dollar industry. The FDA once proposed that cosmetics be regulated as a subcategory under food regulation. Congress failed to enact this proposal in 1906. When Congress enacted the 1938 law, cosmetics became regulated for the first time. Hair-dye toxicity was the motivating force for this regulation. The FDA litigated to gain regulatory uniformity of cosmetics with food and drugs. Many attempts to widen and strengthen cosmetic regulations have found supporters. Thus far, Congress has not been sufficiently persuaded, however, that there are consumer risks warranting broader regulation.

One issue tipped the balance to more regulation in the 1970s. The FDA used the Fair Packaging and Labeling Act and the misbranding provisions of the FD and C Act to enforce cosmetic ingredient labeling and regulations.[3] This requirement has had a significant effect on the cosmetic industry.

Future Regulations

While controversy about the need for wider, more stringent regulation continues, there is an important basic precept about cosmetic regulation.

This single most important precept is that FDA must prove lack of safety. Inherent in this restriction is the importance of the measurement of risk/benefit. Since the FDA will consider no risk as acceptable, questions like, "Is there an acceptable risk for a cosmetic product?" and "How can one measure the benefit of cosmetics?" are basic to its determination.

The FDA limits the benefit in the risk/benefit equation to therapeutic benefits only. They conclude that there is no benefit from cosmetics to balance any risk from the ingredients or the finished products. Industry and others argue that feelings of well-being is a distinct benefit. One could argue that the ability of FDA to protect the cosmetic consumer is compromised by limitations of statutory authority, resource constraints, and scientific uncertainties in the risk/benefit area.

Today's cosmetic development, ingredients, and claims have shifted toward the drug side. This is especially so in claims made for therapeutic benefits such as blemish elimination, wrinkle removal or their prevention, and even some moisturizing claims. Of all the areas that may cause regulations, the issue of "cosmeceuticals" may bring about the most scrutiny we have seen in 50 years.

Changes in FDA's regulatory powers have most often come with crisis episodes. Future crises or highly publicized incidents may continue to be persuasive to legislators. For the present, cosmetic regulation is low-key, low-budget, low-profile, and without current pressure or perception of risks that would expand regulation.

THE LAW—DEFINITIONS

The FD and C Act contains the definition of a cosmetic as "An article or a component of an article which is intended to be used on or in the human body for cleansing, beatifying, promoting attractiveness, or altering the appearance of the user." The two important exemptions in the FD and C Act are for soaps made with traditional ingredients and coal tar hair dyes. Ironically, calls for serious regulation of cosmetics have come because of the exemption of hair dyes when toxicities related to them appeared.

As scientists become more creative and the public becomes more discriminating, the ability to define products as cosmetics or drugs becomes more unclear than ever. These elements come together in products like sunscreens, moisturizers, or retin A in cosmetic products that reduce the damage of the skin.

Drug or Cosmetic?

We should remember that the FD and C Act is a labeling law. The intention of the manufacturer still determines the classification of a product as a drug or a cosmetic. A product labeled for drug use or intended for drug use cannot be a cosmetic. If a drug ingredient and claims are put on the label of a cosmetic product, the entire panoply of drug regulations then apply. One can even imply drug status if one: (1) claims a therapeutic effect from its use; (2) states that use will provide relief from conditions known to require drug action; or (3) implies a change or effect on a bodily function.

With the ever-increasing movement to formulate products that have a targeted effect on certain skin layers, flirting with a drug designation is now common. We are in the "cosmeceuticals" era where there is a blurred distinction between a cosmetic and a pharmaceutical. Whether or not the new technology in cosmetics will produce demands from FDA and Congress to regulate them as drugs is presently unclear.

Perception

When the cosmetic manufacturer uses an ingredient recognized as a drug ingredient in a cosmetic product, the user perceives a drug effect or value and the FDA will likely classify it as a drug. If the label claims of a cosmetic are similar to those of a drug (e.g., alteration of the structure or function of the body), then the FDA will likely regard it as a drug.

Intended Use

The FDA has clarified their position about the drug/cosmetic definitions of a product in a GAO report.[4] "The distinction between a drug and a cosmetic rests upon the intended use of the article. We would have to approach the problem of whether a particular ingredient is a drug on a case-by-case basis. In each case we would have to look at all the facts to determine if we can prove that the product was intended for use as a drug or a cosmetic."

The regulatory demarcation line between drugs and cosmetics has been blurry since the passage of the 1938 Act. There have been many attempts to bring cosmetics under more stringent regulation based on perceived risks from cosmetic ingredients.[5] The dilemma lies in the definitions of both a drug and a cosmetic. These definitions and attempts to rationalize them reminds one of the Alice in Wonderland quote, "When I use a word, it means just what I choose it to mean—neither more nor less."

Physical vs. Physiological

Unknown, but extraordinary amounts of time and effort have been devoted to honing the distinction between drugs and cosmetics.[6,7] The FDA's most recent attempt to show a clear line between these definitions was made in a letter to the industry in response to congressional pressure on the issue. If a product makes a physical claim, it is a cosmetic. However, if there is a physiological claim, it would be a drug. However, as skin science advances and we learn more about actions, many products will be seen to have a physiological effect. As noted, FDA has the flexibility to decide, case-by-case, whether a product is a drug, based on ingredients, claims, and safety.

Safety

Because of the legal definition, the question of the effectiveness of a cosmetic product is rarely the focus of regulatory decisions or enforcement. The issues have usually concerned safety or toxicity problems in several areas: (1) adverse reactions, (2) inherent toxicity of cosmetic ingredients, (3) potential for skin sensitization and/or irritation, (4) illnesses from bacterial contamination, (5) eye effects, including infection, irritation, and possible blindness (6) potential of some cosmetic ingredients for carcinogenity.

Ingredient Reviews

Attention to the toxicity issues has been reflected in the Over-The-Counter (OTC) Drug Review conducted by FDA using expert panels to produce ingredient monographs. Some ingredients in drug products are also in cosmetic products. Examples are hexachlorophene and other antimicrobial ingredients such as preservatives or deodorant ingredients.[8] Some cosmetic formulation ingredients were reviewed both by FDA in the OTC Review and by the CTFA during the Cosmetic Ingredient Review. There were overlapping ingredients in this review mounted jointly by FDA and the CTFA in 1976. A multi-discipline Committee, including dermatologists and toxicologists, reviewed 310 cosmetic ingredients (in 6 years). A series of reports on these ingredients have been issued. The review committee continues to meet and issue reports. This group has been highly praised for their continuing hard work.

The FDA now has the statutory power to contain risks to the public health. It is unlikely that Congress would move for the significant resources needed for regulation unless there is a perceived significant public health risk. Robert P. Brady, in a symposium on new advances in cosmetics, moisturizers, and cutaneous toxicity concluded that a consumer can sort out claims and that the FDA should concentrate on the real issues: whether a product is safe, and if cosmetic products do what they say.[9]

Adulteration

For a person in the cosmetic industry examining regulatory issues, adulteration is surely the most riveting. Adulteration is a legal term of art in the Act (Article 601). It defines the characteristic of a product. If a cosmetic product meets any of the following criteria, it is adulterated.

A cosmetic can be adulterated if it bears or contains a poisonous or deleterious substance, which may render it (temporarily or permanently) injurious to users under "customary or usual," or otherwise labeled, conditions of use. Hair dyes, discussed in the preceding section, are exempt.

A cosmetic that contains or consists of "any filthy, putrid, or decomposed substance"—terms lifted right from food sections of the act—is deemed adulterated.

Cosmetics prepared, packaged, or held under insanitary conditions in which they may have become contaminated or otherwise rendered injurious to health are deemed adulterated. Note that this statement includes only the possibility (may) of becoming contaminated, not the actuality. This grants the FDA inspector considerable latitude to write citations.

A cosmetic is adulterated if its container is composed of a poisonous or deleterious substance that may render the contents injurious to health. This relatively little-used packaging authority makes it possible to control packaging for cosmetics in a manner similar to the Agency's control of packaging of foods.

Finally, adulteration applies to a cosmetic (other than a hair dye) which contains color additives in violation of section 706a of the Act.

COSMETIC LABELING

The Food, Drug, and Cosmetic Act prohibits false and misleading cosmetic labeling or the use of deceptive or improper containers. The labeling requirements, although clear in the enacted 1938 FD and C Act, were not fully implemented until support was found in the prohibition sections of the Fair Packaging and Labeling Act.

A good example of the difficulties in the after-the-fact enforcement occurs in the regulation of misleading cosmetics labeling. FDA has the burden of proof for misleading or false labeling since it cannot be regulated by pre-market review and approval. Special warnings can and have been required for some types of cosmetic products.

Labeling Requirements

A label that conforms to the cosmetic labeling requirements must bear a complete listing of ingredients in legible form and in a prominent location (CFR 21 § 701). Professional cosmetics and free samples may be exempt. Ingredients must be listed in descending order of predominate content. Cosmetics that are also drugs must list the active ingredient first. When the FDA decides that an ingredient has met certain procedures and is a trade secret, the ingredient can be listed as "other ingredients." The specific name of flavors and fragrances can be listed using these generics. Readers should be aware that failure to follow ingredient labeling requirements carefully can incur civil penalties such as seizure and injunction.

Safety Testing

There is no pre-clearance for cosmetics. Thus, the burden is on the manufacturer to adequately show the safety of cosmetic ingredients. This burden is even more apparent with the FDA requiring this statement for cosmetics judged not to have been sufficiently safety tested: "Warning, the safety of this product has not been determined." This statement achieves a strong regulatory impact because its appearance has negative connotations to the consumer. The FDA lacks authority to require safety testing. But the imposition of this warning gives FDA the power to retroactively impose accusation of the failure to warn consumers.

COSMETIC RECORDS AND RECALLS

FDA has limited power to obtain information on cosmetics compared to that obtainable for other product categories. There are no mandatory GMP rules for cosmetics like there are for drugs. The acquisition of information is often obtained indirectly since access to manufacturer's records is not open. Lab samples gathered in the process of inspection are often the basis of adulteration or contamination claims. Other sources are the voluntary information programs dealing with product composition, adverse reaction reports, establishment inspection reports, and voluntary reports from physicians and consumers.

Cosmetic GMPs

In the discussion of cosmetic Good Manufacturing Practices, or rather the lack of them, it becomes clear that the GMP for cosmetic manufacture is rather enigmatic. The producer has a dilemma in determining what level of Good Manufacturing Practice to implement and carry out. It may be acceptable to enforce a general GMP required for regulated products (e.g., food GMPs), but it is not advisable. Most reputable cosmetic manufacturers use the drug GMP regulations as a guide for cosmetic production. This may seem to be over-regulation. However, the major problem areas in cosmetic manufacture are the same as those for drugs. Embracing the drug GMPs will likely prevent expensive remedies later on. Many cosmetic companies have been or are owned by drug companies. Thus, they approach cosmetic manufacture much as they would their drug manufacture.

Companies that do not approach their cosmetic manufacture in this more rigorous fashion often produce an inferior quality product. Implementation of GMPs serves to better protect cosmetic products from adulteration. This caution improves the environmental milieu and permits the carefully selected and tested preservative to work. The testing and selection of raw ingredients, microbial environmental sampling, batch record keeping, quality control sampling, quality assurance auditing, and a sanitary environment improve the final product.

Auditing

In the present regulatory environment, planned, scheduled, and implemented audits may be the best investment a cosmetic producer can make. Normally if a firm has regulatory and quality assurance personnel, they will conduct an in-house or internal audit. The manufacturer must organize this process and conduct the audit as if it were an inspection. Personnel must maintain checklists, records of problems to correct, and a record of corrective actions from the audits.

Most companies rely on annual auditing by internal personnel. One should adjust their timing depending on (1) the results of previous audits, (2) routine quality control checks, and (3) routine audits by in-plant quality assurance. Daily checks and quarterly in-plant audits are needed to keep production within the quality limits. Record keeping and reports of deficiencies noted in the auditing are the basis of the next audit. A report must be done to confirm that corrections have been made. Routine and constant auditing of all aspects of production prevents problems found in the in-depth annual or semi-annual audit. Continuous process improvement should be a daily commitment.

One classifies deficiencies into critical, major, or minor problems. This approach can focus on the areas that need improvement by finding deficiencies that may lead to regulatory action when an inspection occurs.

Because any internal auditing program can be blind to their own defects, an annual external audit is prudent protection for a cosmetics producer. Private consultants exist to supply auditors.

The effects of microbial contamination can be the most serious threat of possible regulatory enforcement for cosmetic products. One should emphasize the specific areas (e.g., plant environment, quality of ingredients, and microbial quality of finished product including preservative effectiveness) causing the major problems in contamination of cosmetics.

Other chapters of this book discuss quality assurance and validation of sanitization of the plant and equipment, cleaning procedures, personal hygiene, and quality of ingredients (see Chapters 3 and 5). Using accepted procedures with documented evidence that the procedures do what they claim to do is the essence of validation.

Recalls

Recalls, seizure, and injunction are the Agency's compliance tools. Recalls of cosmetic products are fairly infrequent, however. When they do occur, the cosmetic manufacturer is often overwhelmed by the work needed to conduct them. Therefore, caution and planning are essential. One should anticipate how to conduct a recall despite having a thoroughly reliable quality system in place. This concept is especially true with respect to microbiological control where the microorganisms are constantly changing due to their evolutionary and adaptational nature.

The enforcement power of FDA is not great for cosmetics. When actions are taken, the FDA most often bases them on adulteration and misbranding. If they base such actions on safety questions, the debate over safety testing may be activated. Most often, safety is questioned based on contamination with potentially pathogenic or toxic microorganisms.

Some steps that the manufacturer may take to avoid this difficult and potentially ruinous action are:

1. Conformity to GMP in manufacturing.
2. Maintenance of current information on safety of all ingredients.
3. Reporting of any significant safety problems to FDA with options for voluntary recall, when required.
4. Maintenance of a recall plan anticipating such an eventuality.
5. Keeping records that permit the manufacturer to mount a recall.
6. Identify legal counsel with specific FDA experience.

CONCLUSION

From a regulatory perspective, adulteration gets at the core of potential enforcement of the FD and C Act. The major approach that the FDA has is to make charges of adulteration and misbranding. Most often, adulteration in cosmetics is the result of microbiological contamination. Considering this, it is interesting that microbiology is often the ugly step-sister to toxicology in a company's safety program.

Microbiological quality, environmental control, product contamination, and spoilage are adulteration charges that can be litigated. Only in rare instances do enforcement options escalate to culminate in a recall. Certainly one wants to avoid a recall at all costs, but one should have a regulatory plan should a recall be needed.

REFERENCES

1. Federal Food, Drug and Cosmetic Act, Title 21.
2. O'Reilly, J. T. 1991. Food and Drug Administration. Regulatory Manual Series. Shepard's/Mcgraw Hill Inc., Colorado Springs.
3. Code of Federal Regulations, Title 21. Part 600 to 799.
4. Govt. Accounting Office (GAO) Report. 1978. "Lack of authority hampers attempts to increased cosmetic safety", HRD78-139 at 131-132.
5. Hayes, A. H. 1983. Address, Commissioner, Food and Drug Administration to the Cosmetics, Toiletry and Fragrance Association, Boca Raton, March 2, 1983.

6a. Weitzman, S. 1969. Drug, Device, Cosmetic? *Food Drug Cosmet. L. J.* 24, 226.

6b. Weitzman, S. 1969. Drug, Device, Cosmetic? Part II. *Food Drug Cosmet. L. J.* 24, 320, 332.

7. Kleinfeld, V. 1967. Cosmetic or Drug. *Food Drug Cosmet. L. J.* 22, 376.
8. Yingling, G. 1978. The effect of the FDA's OTC Drug Review Program on the Cosmetic Industry. *Food Drug Cosmet. L. J.* 33, 78.
9. Brady, R. P. 1991. *Will moisturizers become drugs?* 7th Symposium on Cutaneous Health. Oct. 8-9, 1991, Washington, D.C.

APPENDIX 1

SELECTED REFERENCES FOR COSMETIC MICROBIOLOGY

A BASIC LIBRARY FOR THE COSMETIC MICROBIOLOGIST

Anon., Application of radiation technology to the cosmetic industry, *CTFA Cosmet. J.,* 30, 1983.
Anon., Cosmetic preservation: a comprehensive study of preservatives in oil/water lotions, *Norda Briefs,* 1-3, 1976.
Anon., Cosmetic preservatives encyclopedia, *Cosmet. Toilet.,* 105, 49, 1990.
Anon., Council of Europe lists of anti-perspirants, deodorants and anti-dandruff agents, *Cosmet. Toilet.,* 102, 6, 1987.
Anon., Efficacy of antimicrobial preservation - A proposal for European Pharmacopoeia, *Pharmeuropa,* Feb., 1991.
Anon., *European Community Cosmetics Directive*: Amendments and adaptations to December, 1990 including annex VI preservatives list plus 13th adapting directive, March, 1991.
Anon., Frequency of preservative use in cosmetic formulas as disclosed to FDA - 1990. *Cosmet. Toilet.*, 105, 45, 1990.
Anon., History of cosmetics, *J. Chem. Educ.*, 56, 46, 1979.
Anon., Preservative documentaries, *Cosmet. Toilet.*, 108, 1993.
Anon., Preservative properties of Bronopol, *Cosmet. Toilet.,* 92, 87, 1977.
Anon., Preservatives allowed in Europe, *Cosmet. Toilet.,* 105, 69, 1990.
Anon., Preservatives briefing: Cosmetic preservatives in the UK and Europe - their properties, use, limitations, availability and comment, *Soap Perfum. Cosmet.,* 93, 101, 1982.
Alexander, M., *Microbial Ecology*, John Wiley & Sons, Inc., New York, 1971.
Atlas, R. M. and Bartha, R., *Microbial Ecology: Fundamentals and Applications*, 3rd ed., Benjamin Cummings, Redwood City, CA, 1993.
Atlas, R. M., *Microbiological Media*, 2nd ed., CRC Press, Boca Raton, FL, 1997.

ASTM Standards on Materials and Environmental Microbiology, American Society of Testing and Materials, Philadelphia, PA, 1993.

Balsam, M. S. and Sagarin, E., Eds., *Cosmetics, Science and Technology*, Vol. 3, 2nd edition, Krieger Publishers, Melbourne FL, 1974.

Balows, A. et al., Eds., *Manual of Clinical Microbiology*, 5th ed., American Society for Microbiology, Washington, D.C., 1991.

Bergey's Manual of Systematic Bacteriology, Vol. 1-4, Williams & Wilkins, Baltimore, MD, 1984.

Black, J. G., *Microbiological Principals and Applications*, Prentice Hall, Englewood Cliffs, NJ, 1993.

Block, S. S., Ed., *Disinfection, Sterilization and Preservation*, 4th ed., Lea and Febiger, Philadelphia, 1991.

Bloomfield, S. F., Baird, R., Leck, R. E., and Leech, R., Eds., *Microbiological Quality Assurance in Pharmaceuticals, Cosmetics, and Toiletries*, Ellis Horwood, L. H. Chichester, England, Prentice Hall, Englewood Cliffs, N.J.

Brannan, D. K., Ed., *Cosmetic Microbiology: A Practical Handbook*, CRC Press, Boca Raton, FL, 1997.

Brown, M. R. W. and Gilbert, P., *Microbiological Quality Assurance: A Guide Towards Relevance and Reproducibility of Inocula*, CRC Press, Boca Raton, FL, 1995.

Chakrabarty, A. M., Iglewski, B., Kaplan, S., and Silver, S., Eds., *Evolving Biotechnology*, American Society for Microbiology, Washington, D.C., 1990.

Chapelle, F. H., Ed., *Groundwater Microbiology and Geochemistry*, John Wiley & Sons Inc., New York, 1993.

Characklis, W. G. and Marshall, K. C., Eds., *Biolfilms*, John Wiley & Sons Inc., New York, 1990.

Code of Federal Regulations, Title 21, 1994.

Cohen, Y. and Rosenberg, E., Eds., *Microbial Mats*, American Society for Microbiology, Washington, D.C., 1989.

Compendium of Methods for the Microbiological Examination of Foods, 3rd ed., American Public Health Association, Washington,D.C., 1992.

CTFA Microbiology Guidelines, Cosmetics Toiletries and Fragrance Association, Washington, D.C., 1993.

CTFA Microbiology Committee, *Microtopics (CTFA)*, Allured Publications, Wheaton, IL, 1986.

CTFA Technical Guidelines, Cosmetics Toiletries and Fragrance Association, Washington, D.C., 1985.

FDA Bacteriological Analytical Manual, 6th ed., Association of Official Analytical Chemists, Washington,D.C., 1992.

Franklin, T. J. and Snow, G. A., *Biochemistry of Antimicrobial Action* 4th ed., Chapman and Hall, London, 1989.

Garrett, L., *The Coming Plague*, Penguin Books, New York, 1994.

Gerhardt, P., Murray, R. G. E., Wood, W. A., and Krieg, N. R., *Methods for General and Molecular Bacteriology*, American Society for Microbiology, Washington, D. C., 1994.

Hugo, W. B. and Russell, A. D., *Pharmaceutical Microbiology* 3rd ed., Blackwell Scientific Publications, Oxford, 1983.

Kabara, J. J., Ed., *Cosmetics and Drug Preservation, Principles and Practice*, Marcel Dekker, New York, 1984.

Kabara, J. J. and Orth, D. S., Eds., *Preservative-Free and Self-Preserving Cosmetics and Drugs*, Marcel Dekker, New York, 1997.
Lederberg, J., Ed., *Enclopedia of Microbiology Vol. 1-4*, Academic Press, San Diego, CA, 1992.
The Microbiological Update, Microbiological Applications, Inc., Islamorada, FL.
Mitchell, R., Ed., *Environmental Microbiology* Wiley-Liss, Inc., NY, 1992.
Morpeth, F. F., Ed., *Preservation of Surfactant Formulations*, Blackie Academic, London, 1995.
Noble, W. C., *Microbiology of Human Skin*, 2nd ed., Lloyd-Luke Ltd., London, 1981.
Official Methods of Analysis of the AOAC, 16th ed., Association of Official Analytical Chemists, Washington, D.C., 1994.
O'Leary, W. M., Ed., *Practical Handbook of Microbiology*, CRC Press, Boca Raton, FL., 1989.
Orth, D. S., *Handbook of Cosmetic Microbiology*, Marcel Dekker, NY, 1993
Poindexter, J. S. and Leadbetter, E. R., Eds., *Bacteria in Nature, Vol. 2, Methods and Special Applications in Bacterial Ecology*, Plenum Press, New York, 1986.
Russell, A. D., Hugo, W. B., and Ayliffe, G. A. J., *Disinfection, Preservation, and Sterilization,* 2nd. ed., Blackwell Scientific Publications, Oxford, 1992.
Sonea, S. and Panisset, M. A New Bacteriology, Boston, MA:Jones & Bartlett, 1983.
Standard Methods for the Examination of Dairy Products, 16th ed., American Public Health Association, Washington,D.C., 1993.
Standard Methods for the Examination of Water and Waste Water, American Public Health Association, Washington,D.C., 1992.
Troller, J. A., *Sanitation in Food Processing*, Academic Press, Orlando, FL, 1983.
Troller, J. A. and Christian, J. H. B., Eds., *Water Activity and Food*, Academic Press, New York, 1978.
U.S. Pharmacopeia XXIII, U.S. Pharmacopeial Convention, Rockville, MD, 1995.

STANDARD LITERATURE REFERENCES FOR THE COSMETIC MICROBIOLOGIST

Ahearn, D. G. and Wilson, L. A., Microflora of the outer eye and eye area cosmetics, *Dev. Ind. Microbiol.,* 17, 23, 1976.
Ahearn, D. G., Wilson, L. A., Julian, A. J., Reinhardt, D. J., and Ajello, G., Microbial growth in eye cosmetics: Contamination during use, *Dev. Ind. Microbiol.,* 15, 211, 1974.
Akimoto, N. et al., Determination of p-hydroxybenzoic acid esters in cosmetics by liquid chromatography with ultraviolet and fluorescence detection, *J. Off. Anal. Chem.*, 71, 823, 1988.
Alexander, P., Preservatives and stabilizers in the patent literature 1982-87, *Cosmet. Toilet.*, 102, 89, 1987.
Alexander, P., Cosmetics creation, *Manuf. Chem.,* 61, 24, 1990.
Alguire, D., Microorganism control of cosmetic ingredients with 100% ethylene oxide, *Amer. Perfum. Cosmet.,* 85, 31, 1970.
Alguire, D. E. and Yeung, A. C., Making cosmetics microbiologically safe, *Cosmet. Toilet.*, 94, 77, 1979.

Anderson, D. W. and Ayers, M., Microbiological profile of selected cosmetic products with and without preservatives after use, *J. Soc. Cosmet. Chem.*, 23, 863, 1972.
Anderson, D. W., McConville, J. F., and Anger, C. B., Microbiological profile of used eye cosmetics by examination of product only, *Cosmet. Perfum.*, 88, 25, 1973.
Anderson, D. W., McConville, J. F., and Anger, C. B., Some comments on the microbiological profile of used automatic eye cosmetics by examination of both applicator and product, *Cosmet. Perfum.*, 88, 29, 1973.
Asano, A. and Faust, R. E., Evaluation of cosmetically-useful antimicrobials, *Amer. Perfum. Cosmet.*, 81, 55, 1966.
Ashour, M. S. E., Hefnai, H., El-Tayeb, O. M., and Abdelaziz, A. A., Microbial contamination of cosmetics and personal care items in Egypt, *Cosmet.Toilet.*, 102, 61, 1987.
Ashworth, D. and Gutherie, W., A new concept in protecting against spoilage microflora, in *Cosmetics and Toiletries Manufacture Worldwide*, Aston Press, Hertfordshire, England, 1993.
Ashworth, D. and Turton, K., An alternative to removing preservatives from formulations, in *Cosmetics and Toiletries Manufacture Worldwide*, Aston Press, Hertfordshire, England, 1995.
Bachrach, E. E., How you can control the FDA cosmetic inspection, *CTFA Cosmet. J.*, 22, 1983.
Baird, R. M., Microbial contamiation of cosmetic products, *J. Soc. Cosmet. Chem.* 28, 17, 1977.
Baird, R. M., Bacteriological contamination of products used for skin care in babies, *Int. J. Cosmet. Sci.* 6, 85, 1984.
Bandelin, F. J., Antibacterial and preservative properties of alcohols, *Cosmet. Toilet.*, 92, 1977.
Bandelin, F. J., The effect of pH on the efficiency of various mold-inhibiting compounds, *J. Am. Pharm Assoc. Sci. Ed.*, 46, 691, 1959.
Barnes, M. and Denton, G. W., Capacity tests for the evaluation of preservatives in formulations, *Soap Perf. Cosmet.*, 42, 729, 1969.
Beach, W. J., Production/engineering: Good microbiological manufacturing practices for aqueous cosmetics and toiletries, *Soap Chem. Spec.*, 45, 144, 1969.
Beach, W. J., Production/engineering: microbial content of cosmetics and toiletries, *Soap Chem. Spec.*, 45, 142, 1969.
Beach, W. J., Production/engineering, microbial contamination of cosmetics, *Soap Chem. Spec.*, 46, 156, 1970.
Bean, H. S., Preservatives for pharmeceuticals, *J. Soc. Cosmet. Chem.*, 23, 703, 1972.
Bean, H. S. and Herman-Ackah, S. M., Influence of oil:water ratio on the activity of some bactericides against *Escherichia coli* in liquid paraffin and water dispersions, *J. Pharm. Pharmacol.*, 16, 58T, 1964.
Bean, H. S., Konnig, G. H., and Malcolm, S. A., A model for the influence of emulsion formulations on the activity of phenolic preservatives, *J. Pharm. Pharmacol.*, 21, 173S, 1969.
Bean, H. S., Konnig, G. H., and Thomas, J., Significance of the partition coefficient of a preservative in cosmetic emulsions, *Am. Perf. Cosmet.*, 85, 61, 1970.
Beckett, A. H. and Robinson, A. E., The inactivation of preservatives by nonionic surface active agents, *Soap Perf. Cosmet.*, 31, 454, 1958

Beerens, H., Romond, C., and Lemaire, D., Techniques for the enumeration of microorganisms in cosmetic preparations, *Soap Perf. Cosmet.*, 48, 262, 1976.

Bell, K. W., Some aspects of microbiology in cosmetic products, *Amer. Perf. Cosmet.*, 81, 25, 1966.

Benassi, C. A., Dehydroacetic acid sodium salt stability in cosmetic preservative mixtures, *Int. J. Cosmet. Sci.*, 10, 29, 1988.

Benassi, C. A., Qualitative and Quantitative determination of prevan in cosmetic emulsions by direct analysis via fast atom bombardment/collisional spectroscopy, *Biomed. Environ. Mass Spectrom.*, 18, 855, 1989.

Benassi, C. A. et al., Chemical stability and microbiological reply of preservative systems in cosmetics: a direct correlation, *Int. J. Cosmet. Sci.*, 10, 231, 1988.

Benassi, C. A. et al., Interaction between dehydroacetic acid sodium salt and formaldehyde: structural identification of the product, *J. Soc. Cosmet. Chem.*, 39, 85, 1988.

Benassi, C. A. et al., High-performance liquid chromatographic determination of free formaldehyde in cosmetics preserved with Dowicil 200, *J. Chromatogr.*, 502, 193, 1990.

Benassi, C. A., Rettore, A., Semenzato, A., Bettero, A., and Cerini, R., Chemical stability and microbiological reply of preservative systems in cosmetics: A direct correlation, *Int. J. Cosmet. Sci.*, 10, 231, 1988.

Berke, P. A. and Rosen, W. E., Germall, a new family of antimicrobial preservatives for cosmetics, *Amer. Perfum. Cosmet.*, 85, 1970.

Berke, P. A. and Rosen, W. E., Imidazolidinyl urea activiity against *Pseudomonas*, *J. Soc. Cosmet. Chem.*, 29, 757, 1978.

Berke, P. A. and Rosen, W. E., Are cosmetic emulsions adequately preserved against *Pseudomonas*, *J. Soc. Cosmet. Chem.*, 31, 37, 1980.

Berke, P. A., Steinberg, D. C., and Rosen, W. E., Germaben II: A complete preservative system in clear liquid form, *Cosmet. Toilet.*, 97, 89, 1982.

Bettero, A. et al., The characterization of isothiazolinone preservatives in cosmetics, *J. Pharm. Biomed. Anal.*, 3, 581, 1985.

Bhadauria, R. and Ahearn, D. G., Loss of effectiveness of preservative systems of mascaras with age, *Appl. Environ. Microbiol.*, 39, 665, 1980.

Bhargava, H. N. and Anaebonam, A., Essentials of cosmetic preservation, *Soap Cosmet. Chem. Spec.*, 59, 39, 1983.

Blachman, U. and Elowitz-Jeffes, L., Microbiology of cosmetics - regulatory and quality-assurance aspects, *Cosmet. Tech.*, Jan. 24, 1982.

Blakeway, J. M., Fragrances as preservatives, *Seifen Oele Fette Wachse*, 116, 357, 1990.

Blaug, S. M. and Ahsan, S. S., Interaction of sorbic acid with nonionic macromolecules, *J. Pharm. Sci.*, 50, 138, 1961.

Boehm, E. E. and Maddox, D. N., Problems of cosmetic preservation, *Manufac. Chem. Aerosol News*, 42, 41, 1971.

Bonorden, R., The nature and extent of bacteria found in some raw materials used in the cosmetic industry, *CTFA Cosmet. J.*, 5, 23, 1973.

Borovian, G. E., *Pseudomonas cepacia*: Growth in and adaptability to increased preservative concentrations, *J. Soc. Cosmet. Chem.*, 34, 197, 1983.

Bowman, P. I. and Lindstrom, S. M., Resistance of *Pseudomonas* strains to imidazolidinyl urea, *J. Soc. Cosmet. Chem.*, 36, 413, 1985.

Branna, T., Preservatives go global, *HAPPI*, May, 1994.

Brannan, D. K., Cosmetic Microbiology, in *Encyclopedia of Microbiology*, Lederberg, J., Ed., Academic Press, San Diego, CA, 1992, 593.

Brannan, D. K., Cosmetic preservation, *J. Soc. Cosmet. Chem.*, 46, 199, 1995.

Brannan, D. K., Preservation of personal care products, in *Preservation of Surfactant Formulations*, Morpeth, F., Ed., Black Academic, London, 1995.

Brannan, D. K., The role of packaging in product preservation, in *Preservative-Free and Self-Preserving Cosmetics and Drug Products*, Kabara, J. J. and Orth, D. S., Marcel Dekker, New York, 1997, chap. 10.

Brannan, D. K., Cosmetic preservation, *Cosmet. Toilet.*, 111, 69, 1996.

Brannan, D. K., Dille, J. C., and Kaufman, D. J., Correlation of *in vitro* challenge testing with consumer-use testing for cosmetic products, *Appl. Environ. Microbiol.*, 53, 1827, 1987.

Brannan, D. K. and Dille, J. C., Type of closure prevents microbial contamination of cosmetics during consumer use, *Appl. Environ. Microbiol.*, 56, 1476, 1990.

Breach, G. D., Microbiological quality control a case history, *J. Soc. Cosmet. Chem.*, 26, 315, 1975.

Breach, G. D., DeNavarre, M. G., Idson, B., Kanig, J. L., and Shansky, A., Fingers or applicators (prevention of microbial contamination in eye cosmetics), *Cosmet. Perfum.*, 88, 53, 1973.

Bruch, C. W., Control of microorganisms in cosmetic products, *Amer. Perfum. Cosmet.*, 86, 47, 1971.

Bruch, C. W., Cosmetics: sterility vs. microbial control, *Amer. Perfum. Cosmet.*, 86, 45, 1971.

Bruch, C. W., Microbiological quality of topical products, *Drug Cosmet. Ind.*, 109, 26, 1971.

Bruch, C. W., Possible modifications of USP--microbial limits and tests, *Drug Cosmet. Ind.*, 110, 32, 1972.

Bruch, C. W., Objectionable microorganisms in nonsterile drugs and cosmetics, *Drug Cosmet. Ind.*, 11, 51, 1972.

Bruch, C. W., Eye products: handle with care, *FDA Consum.*, 6, 5, 1972.

Bruze, M. et al., Contact allergy to the active ingredients of Kathon CG, *Cont. Dermat.*, 16, 183, 1987.

Bruze, M. et al., Contact allergy to the active ingredients of Kathon CG in the guinea pig, *Acta Derm. Venerol.*, 67, 1987.

Bryan, F. L., Hazard analysis and critical control point (HACCP) concept, in *Dairy, Food, and Environmental Sanitation*, 1990.

Bryce, D. M., Quality in a commercial environment, *Cosmet. Toilet.*, 105, 83, 1990.

Buchanan, J. A. and Philp, R., Simultaneous HPLC determination of Dowicil 200 and methyl- and propyl-parabens in aqueous-based mascaras and cosmetic powders by the internal standard method, *LC-GC*, 5, 52, 1987.

Bühlmann, X., Method for microbiological testing of nonsterile pharmaceuticals. *Appl. Microbiol.*, 16, 1919, 1968.

Butler, N. and McCarthy, T. J., Parameters of a dialysis method for studying the release of preservatives, *Cosmet. Toilet.*, 95, 47, 1980.

Canby, T. Y., Bacteria ... teaching old bugs new tricks, *Nat. Geograph.*, Aug., 1993.

Chan, M. and Bruce, H. N., A rapid screening test for ranking preservative efficacy, *Drug Cosmet. Ind.*, 135, 34, 1981.

Chapman, D. G., Preservatives available for use, in *Society for Applied Bacteriology Technical Series No. 22, Preservatives in the Food, Pharmaceutical and Environmental Industries*, Board, R.G., Allwood, M.C., and Banks, J. G., Eds., Blackwell Scientific Publications, Edinburgh, 1987, 177.

Charig, A. et al., Inhibitor of odor-producing axillary bacterial exoenzymes, *J. Soc. Cosmet. Chem.*, 42, 133, 1991.
Charles, R. D. and Carter, P. J., The effect of sorbic acid and other preservatives on organism growth in typical nonionic emulsified commercial cosmetics, *J. Soc. Cosmet. Chem.*, 42, 383, 1991.
Christensen, S. G., The microflora in cosmetics and nonsterile pharmaceuticals, and the microbiological standards for these products, *J. Nordisk Medici. (Stockholm)*, 94, 165, 1979.
Chu, P. K. M., The efficacy of some commonly used preservatives in cosmetics products, M. S. Dissertation, California State University, Long Beach, 1974.
Coates, D., Interaction between preservatives, plastics, and rubbers, *Mfg. Chem. Aerosol News*, 44, 19, 1973.
Coates, D., Interaction between preservatives and surfactants, *Mfg. Chem. Aerosol News*, 44, 41, 1973.
Coates, D., Preservative/colloid interaction, *Mfg. Chem. Aerosol News*, 44, 34, 1973.
Coates, D. and Richardson, G., Relations between estimates of binding of antimicrobial agents by macromolecules, based on physiochemical and microbiological data, cetylpyridinium chloride and polyethylene glycol, *J. Appl. Bacteriol.*, 36, 249, 1973.
Coates, D. and Woodford, R., Methods available for studying antimicrobial interaction in cosmetics, *Cosm. Perfum.*, 88, 43, 1973.
Cooper, M. S., Preservative efficacy: Compedial and regulatory issues, *J. Parenter. Sci. Tech.*, 43, 187, 1989.
Cooper, M. S., Antimicrobial preservation, in *Microbiological Update*, Islamorada, FL, 1994-1995.
Cosmetic, Toiletry, and Fragrance Association, Final report on the safety assessment of methylparaben, ethylparaben, propylparaben and butylparaben, *J. Am. Coll. Toxicol.*, 3, 147, 1984.
Cosmetic, Toiletry, and Fragrance Association, Addendum to final report on safety assessment of 2-bromo-2-nitropropane-1,3-diol., *J. Am. Coll. Toxicol.*, 3, 139, 1984.
Cosmetic, Toiletry, and Fragrance Association, Final report on the safety assessment of quaternium-15, *J. Am. Coll. Toxicol.*, 5, 61, 1986.
Cosmetic, Toiletry, and Fragrance Association, Final report on the safety assessment of DMDM hydatoin, *J. Am. Coll. Toxicol.*, 7, 245, 1988.
Cowen, R. A. and Steiger, B., Antimicrobial activity—A critical review of test methods of preservative efficacy, *J. Soc. Cosmet. Chem.*, 27, 67, 1976.
Cowen, R. A. and Steiger, B., Why a preservative system must be tailored to a specific product, *Cosmet. Toilet.*, 92, 15, 1977.
Croshaw, B., Preservatives for cosmetics and toiletries, *J. Soc. Cosm. Chem.*, 28, 3, 1977.
Croshaw, B., Preservation of emolient and moisturizing preparations: Effect of partition coefficients, *Cosm. Toilet.*, 93, 42, 1978.
CTFA Microbial Content Subcommittee, CTFA national microbiological survey of cosmetics and toiletries 1972-1975, *CTFA Cosmet. J.*, 9, 24, 1977.
CTFA Microbial Content Subcommittee, Microbiological limit guidelines for cosmetics and toiletries, *CTFA Cosmet. J.*, 4, 25, 1972.
CTFA Microbiology Quality Assurance Subcommitee, Microbiological guidelines for process water, in *CTFA Technical Guidelines*, Cosmetic, Toiletry, and Fragrance Association, Washington, D.C., 1981.

CFTA Quality Assurance Subcommittee, Facility housekeeping, Unpublished paper.

CTFA Microbiology Quality Assurance Committee, Microbiological sampling guideline for the cosmetic industry, in *CTFA Technical Guidelines*, Cosmetic, Toiletry, and Fragrance Association, Washington, D.C., 1981.

CTFA Microbiology Quality Assurance Subcommittee, Microbiological aspects of quality assurance, in *CTFA Technical Guidelines*, Cosmetic, Toiletry, and Fragrance Association, Washington, D.C., 1981.

CTFA Preservation Subcommittee, A study of the use of rechallenge in preservative testing of cosmetics, *CTFA Cosmet. J.*, 13, 19, 1981.

CTFA Preservation Subcommittee, A guideline for the determination of adequacy of preservation of cosmetics and toiletry formulations, in *CTFA Technical Guidelines*, Cosmetic, Toiletry, and Fragrance Association, Washington, D.C., 1981.

CTFA Preservation Subcommittee, A guideline for preservation testing of aqueous liquid and semi-liquid eye cosmetics, in *CTFA Technical Guidelines*, Cosmetic, Toiletry, and Fragrance Association, Washington, D.C., 1981.

CFTA Cosmetic Ingredient Dictionary, Cosmetic, Toiletry, and Fragrance Association, Washington, D. C., 1993.

CTFA Cosmetic Ingredient Handbook, Cosmetic, Toiletry, and Fragrance Association, Washington, D. C., 1993.

CTFA Microbiology Committee, CTFA survey: Preservative test methods companies use, *Cosmet. Toilet.*, 105, 79, 1990.

Curry, J. C., A history of cosmetic microbiology, *Cosmet. Toilet.*, 98, 1983.

Curry, J. C. and Butera, E. G., A laboratory comparison of selective media for pseudomonas in cosmetics and toiletries, *Dev. Ind. Microbiol.*, 12, 165, 1971.

Curry, J. C., Bactericidal activity laboratory test, *Soap Chem. Spec.*, Mar, 1968.

Curry, J. C., Cosmetic microbiology 50 years of change—An historical perspective of microbiology and cosmetics, *HAPPI*, 8, 44, 1983.

Curry, J. C., Gradient plate bioassay for tyrothricin and its application to dentifrices, *Appl. Microbiol.*, 11, 539, 1963.

Curry, J. C., Manipulation of water activity for developing products intrinsically hostile to microbial growth, in *Cosmetics and Toiletries Manufacture Worldwide*, Aston Press, Hertfordshire England, 1995.

Curry, J. C., Microbiological aspects of cosmetics and toiletries manufacturing, in *Cosmetics and Toiletries Manufacture Worldwide*, Aston Press, Hertfordshire England, 1993.

Curry, J. C., The gradient plate procedure for rapid screening of antibacterials, presented at Chemical Specialties Manufacturers Association, 52nd Annual Meeting Proceedings, 1965.

Curry, J. C., Thoughts on preservation testing of water-based products, *Cosmet. Toilet.*, 102, 93, 1987.

Curry, J. C., Water activity and preservation, *Cosmet. Toilet.*, 100, 53, 1985.

Davis, D. A., Preservatives, *Drug Cosmet. Ind.*, Dec., 1994.

Davis, J. G., Fundamentals of microbiology in relation to cleansing in the cosmetic industry, *J. Soc. Cosmet. Chem.*, 23, 45, 1972.

Davis, J. G., Microbiological stability of cosmetic and toilet preparations, *Soap Perf. Cosm.*, 46, 409, 1973.

Davis, J. G., Microbiological stability and hazards of cosmetic preparations, *Soap Perf. Cosm.*, 53, 133, 1980.

Davison, A. L., The validity of the criteria of pharmacopeial preservative efficacy tests. (British) *Pharmaceutical J.*, May, 1991.

Dawson, F. W., Some techniques for microbial control in manufacturing plants. *J. Soc. Cosmet. Chem.*, 24, 655, 1973.

Dawson, N. L. and Reinhardt, D. J., Microbial flora of in-use, display eye shadow testers and bacterial challenges of unused eye shadows, *Appl. Environ. Microbiol.*, 42, 297, 1981.

Decker, R. L., Jr. and Wenninger, J.A., Frequency of preservative use in cosmetic formulas as disclosed to FDA - 1987, *Cosmet. Toilet.*, 102, 21, 1987.

DeGroot, A. C., Kathon CG: Cosmetic allergy and patch test sensitization. *Cont. Dermat.*, 12, 76, 1985.

DeGroot, A. C., Kathon CG: A review, *J. Am. Acad. Dermat.*, 18, 350, 1988.

DeGroot, A. C., Isothiazolinone preservative as important contact allergen in cosmetics, *Derm. Beruf. Umwelt.*, 35, 169, 1987.

DeGroot, A. C. et al., Study of the interaction of Kathon CG-Germall II in hydrophilic creams. *Int. J. Cosmet. Sci.*, 9, 49, 1987.

DeKruijf, N., Determination of preservatives in cosmetic products, II. High-performance liquid chromatographic identification, *J. Chromatogr.*, 469, 317, 1989.

DeKruijf, N. et al., Determination of preservatives in cosmetic products, I. Thin-layer chromotographic procedure for the identification of preservatives in cosmetic products, *J. Chromatogr.*, 410, 395, 1987.

Denyer, S. P. et al., Synergy in preservative combinations, *Int. J. Pharm.*, 25, 345, 1985.

Diehl, K. H., Important secondary conditions for optimal use of cosmetic preservatives, *Parfeum. Kosmet.*, 71, 396, 1990.

Dixon, H. and Dolan, C. S., Occurrence of mercury-resistant *Pseudomonas* in emulsion formulations preserved with phenylmercuric acetate, *Int. J. Cosmet. Sci.*, 3, 261, 1981.

Dony, J., Microbiological problems of cosmetics, *J. Pharm. Belg.*, 30, 233, 1975.

Duc, T. L., Concentrated solutions of preservatives: Their technological and analytical advantages in cosmetic products, *Int. J. Cosmet. Sci.*, 5, 9, 1983.

Dunnigan, A. P., Microbiological control of cosmetic products, in *Proc. Joint Conf. for Cosmetic Science*, Toilet Goods Association, Washington, D. C., 1968, 179.

Dunnigan, A. P. and Evans, J. R., Proposed methodology for the isolation of gram negative microorganisms from ointments and lotions, *TGA Cosmet. J.*, 38, 1969.

Dunnigan, D. A. and Evans, J. R., Report of a special survey: Microbiological contamination of topical drugs and cosmetics, *TGA Cosmet. J.*, 2, 39, 1970.

Durant, C. and Higdon, P., Preservation of cosmetic and toiletry products, in *Soc. Appl. Bacteriol. Tech. Ser. no. 22, Preservatives in the Food, Pharmaceutical and Environmental Industries*, Board, R. G. et al., Eds., Blackwell Scientific, Edinburgh, 1987, 231.

Durant, C. and Higdon, P., Methods for assessing antimicrobial activity, in *Soc. Appl. Bacteriol. Tech. Ser. 27, Mechanisms of Action of Chemical Biocides*, Denyer, S. P. and Hugo, W. B., Eds., Blackwell Scientific Publications, 1991.

Eachus, A. C. et al., Oxaban A—An alternative preservative for cosmetics and toiletries, *Seifen Oele Fette Wachse*, 116, 537, 1990.

Edwards, C. C., Use of mercury in cosmetics including use as skin-bleaching agent in cosmetic preparations also regarded as drugs, (proposed rule making) CFR 21, 12967, 1972.

Eiermann, H. J., FDA regulation of cosmetics, *Cosmet.Toilet.*, 98, 1983.

Eiermann, H. J., Contribution of the Microbiologist to Cosmetic product Safety, *Drug Cosmet. Ind.*, 119, 43, 1976.

Eiermann, H. J., Cosmetic Regulatory Activities in the United States: Past, Present and Future, *Cont. Dermat.*, 4, 157, 1978.

Engley, F. B. and Dey, B. P., A universal neutralizing medium for antimicrobial chemicals, in *Proc. of the 56th Mid-Year Meeting of the Chemical Specialties Manufacturers Association*, New York, 1970, 100.

Entrekin, D. N., Relation of pH to preservative effectiveness, *Int. J. Pharm. Sci.*, 50, 743, 1961.

Evans, J. R., Gilden, M. M., and Bruch, C. W., Methods for Isolating and identifying objectionable gram negative bacteria and endotoxins from topical products, *J. Soc. Cosmet. Chem.*, 23, 549, 1972.

Evans, W. P., The solubilization and inactivation of preservatives by non-ionic detergents, *J. Pharm. Pharmacol.*, 16, 323, 1964.

Facino, R. M. et al., Identification of preservative composition in finished cosmetic formulations by collisionally activated decomposition mass-analyzed ion kinetic energy spectrometry, *Biomed. Environ. Mass Spectrom.*, 19, 493, 1990.

Farrington, J. K., Martz, E. L., Wells, S. J., Ennis, C. C., Holder, J., Levchuk, J. W., Avis, K. E., Hoffman, P. S., Hitchins, A. D., and Madden, J. M., Ability of laboratory methods to predict in-use efficacy of antimicrobial preservatives in an experimental cosmetic, *Appl. Environ. Microbiol.*, 60, 4553, 1994.

Faverro, M. S., *Pseudomonas aeruginosa*—Growth in distilled water from hospitals, *Science*, 173, 836, 1971.

Favet, J. et al., Adaptation of *Escherichia coli, Pseudomonas aeruginosa* and *Staphylococcus aureus* to Kathon CG and Germall II in an O/W cream, *Cosmet. Toilet.*, 102, 75, 1987.

Fenn, R. J. and Csejka, D. A., The stability of 2-pyridinethiol-1-oxide sodium salt as a function of pH, *J. Soc. Cosmet. Chem.*, 33, 243, 1982.

Flawn, P. C., Malcolm, S. A. and Woodroffe, R. C. S. Assessment of the preservative capacity of shampoos, *J. Soc. Cosmet. Chem.*, 24, 229, 1973.

Foussereau, J., An epidemiological study of contact alergy to 5-chloro-3-methyl isothiazolone/3-methyl isothiazolone, in *Strasbourg Cont. Dermat.*, 22, 68, 1990.

Frank, C. L., Microbiological contamination in industry, *Amer. Perfum. Cosmet.*, 84, 49, 1969.

Frosch, P. J. et al., Contact allergy to Bronopol, *Cont. Dermat.*, 22, 24, 1990.

Gagliardi, L. et al., Determination of aromatic alcohols in cosmetic products using reversed-phase high-performance liquid chromatography, *J. Chromatogr.*, 294, 442, 1984.

Gagliardi, L. et al., Determination of phenolic preservatives in cosmetic products using reversed-phase high-performance liquid chromatography, *Farmaco [Prat]*, 40, 165, 1985.

Gagliardi, L. et al., Determination of preservatives in cosmetic products by reversed-phased high-performance liquid chromatography I, *J. Chromatogr.*, 315, 465, 1984.

Gagliardi, L. et al., Determination of preservatives in cosmetic products by reversed-phase high-performance liquid chromatography II, *J. Chromatogr.*, 325, 353, 1985.

Gagliardi, L. et al., Determination of preservatives in cosmetic products by ion-pair reversed-phase high-performance liquid chromatography III, *J. Chromatogr.*, 348, 321, 1985.

Gagliardi, L. et al., Determination of preservatives in cosmetic products by reversed-phase high-performance chromatography IV, *J. Chromatogr.*, 508, 252, 1990.

Geis, P. A., Preservation of cosmetics and consumer products: Rationale and application, *Dev. Ind. Microbiol.*, 29, 305, 1988.

Geldsetzer, K., Preservative efficacy test in national pharmacopeias [British, Italian, German, French], *Pharmacology* (German), Feb, 1990.

Gershenfield, L., Antimicrobial agents in cosmetic preparations, *Amer. Perf. Cosm.*, 78, 55, 1963.

Goldman, C. L., Microbiological tests used in a quality control program, *Cosmet. Perf.*, 88, 39, 1973.

Goldman, C. L., Microorganisms isolated from cosmetics, *Drug Cosmet. Ind.*, 117, 40, 1975.

Goldman, C. L., A microbiological case history of a cosmetic product, *TGA Cosmet. J.*, 1, 42, 1969.

Graeveni, A., Pathogenic significance of *P. fluorescens* and *P. putida*, *Yale J. Biol.*, 44, 265, 1971.

Grant, P., Rapid detection of microorganisms using bioluminescence—A brief review, in *Cosmetics and Toiletries Manufacture Worldwide*, Aston Press, Hertfordshire England, 1995.

Gucklhorn, I. R., Cosmetic microbiology, *Mfg. Chem.*, 39, 23, 1968.

Gucklhorn, I. R., Antimicrobials in cosmetics—Tego ampholytic surfactants and betaines, *Mfg. Chem.*, 40, 43, 1969.

Gucklhorn, I. R., Antimicrobials in cosmetics, Part I., *Mfg. Chem. Aerosol News*, 40, 23, 1969.

Gucklhorn, I. R., Antimicrobials in cosmetics second series, *Mfg. Chem. Aerosol News*, 41, 6, 1970 (and the next 8 succeeding issues).

Hall, A. L., Phenoxyethanol: A cosmetically acceptable preservative, *Cosmet. Toilet.*, 96, 83, 1981.

Halleck, F. E., A guideline for the determination of adequacy of preservation of toiletry and cosmetic preparations, *TGA Cosmet. J.*, 2, 20, 1970.

Halleck, F. E., Microbiological problems associated with the development and production of cosmetic and toiletry products, *SIM Dev. Indust. Micro,* 12, 155, 1971.

Hart, J. R., EDTA-type chelating agents in personal care products, *Cosmet. Toilet.*, 98, 54, 1983.

Hasson, A. et al., Patch test sensitivity to preservative Kathon CG® in Spain, *Cont. Dermat.*, 22, 257, 1990.

Henry, S. M. and Jacobs, G., Preservatives documentary: cosmetic preservatives, *Cosmet. Toilet.*, 96, 29, 1981.

Henry, S. M. and Jacobs, G., Cosmetic preservatives—1982 update, *Cosmet. Toilet.*, 97, 36, 1982.

Herman, L. G., The slow growing pigmented water bacteria: problems and sources, *Adv. Appl. Microbiol.*, 23, 155, 1978.

Hiller, J. J., Formulating for hard water, *Cosmet. Toilet.*, 98, 59, 1983.

Horn, N. R., McCarthy, T. J. K., and Price, C. H., Interaction between preservatives and suspension systems, *Am. Perf. Cosmet.*, 86, 37, 1971.

Hugenholtz, P. and Fuerst, J. A., Heterotrophic bacteria in an air-handling system, *Appl. Environ. Microbiol.*, 58, 3914, 1992.
Imbus, H. R., Clinical evaluation of patients with complaints related to farmaldehyde exposure, *J. Allergy Clin. Immunol.*, 76, 831, 1985.
Jacobs, G., Henry, S. M., and Cotty, V. F., The influence of pH, emulsifier and accelerated aging upon preservative requirements of O/W emulsions, *J. Soc. Cosmet. Chem.*, 26, 105, 1975.
Janik, D., Hall, C. S., and DeNavarre, M. G., Glutaraldehyde - a sanitizing agent for the equipment used in the manufacture of cosmetics, *Cosmet. Toilet.*, 92, 99, 1977.
Jarvis, B., Reynolds, A. J., Rhodes, A. C., and Armstrong, M., Survey of microbiological contamination in cosmetics and toiletries in the U.K., *J. Soc. Cosmet. Chem.*, 35, 563, 1974.
Jasnow, S. B. and Smith, J. L., Microwave sanitization of color additives used in cosmetics: Feasibility study, *Appl. Micro.*, 30, 205, 1975.
Jass, H. E., History and status of formaldehyde in cosmetics industry. *Adv. Chem. Ser.*, 210, 229, 1985.
Jayasekara, B. and Claydon, E. J., Determination of low level microbiological contamination in cosmetics by thin film agar, *Cosmet. Toilet*, 93, 67, 1978.
Jermini, M. F. G. and Schmidt-Lorenz, W., Activity of sodium benzoate and ethyl paraben against osmotolerant yeasts at different water activity values, *J. Food. Prot.*, 1987.
Judas, D. M., Microbiological testing of cosmetics, *Drug Cosmet.*, 109, 48, 1971.
Kabara, J. J., Lipids as safe and effective antimicrobial agents for cosmetics and pharmaceuticals, *Cosmet. Perfume*, 90, 21, 1975.
Kabara, J. J., Multi-functional food-grade preservatives in cosmetics, *Drug Cosmet.*, 125:60, 1979.
Kabara: J. J., GRAS antimicrobial agents for cosmetic products, *J. Soc. Cosmet. Chem.*, 31, 1, 1980.
Kabara, J. J., The medium is the preservative, *Cosmet. Toilet.*, 96, 63, 1981.
Kabara, J. J., Cosmetic formulas preserved with food grade chemicals, *Cosmet. Toilet.*, 97, 77, 1982.
Kaiserman, J. M., Moral, J., and Wolf, B. A., A rapid impedimetric procedure to determine bacterial content in cosmetic formulation, *J. Soc. Cosmet. Chem.*, 40, 21, 1989.
Kalling, L. O. et al., Microbiological contamination of medical preparations, *Acta Pharm. Suecica*, 3, 291, 1969.
Kamm, R. E., Some legal aspects of cosmetic preservation, *Cosmet. Toilet.*, 92, 75, 1977.
Kano, C., Nakata, O., Kurosaki, S., and Yanagi, M., Microbial quality control for the manufacture of cosmetic emulsions, *J. Soc. Cosmet. Chem.*, 27, 73, 1976.
Kantor, G. R. et al., Acute allergic contact dermatitis from diazolidinyl urea (Germall II) in a hair gel, *J. Am. Acad. Dermatol.*, 13, 116, 1985.
Kazmi, J. A. and Mitchell, A. G., The inactivation of preservatives, *Soap Perf. Cosm.*, 45, 549, 1972.
Kenney, D., Cosmetic formulas preserved with food grade chemicals, *Cosmet. Toilet.*, 97, 71, 1982.
Korting, H. C. et al., Changes in skin pH and resident flora by washing with synthetic detergent compositions at pH 5.5 and 8.5, *J. Soc. Cosmet. Chem.*, 43, 1991.

Koumanova, R. et al., Evaluating Bronopol, *Mfg. Chem.*, 60, 36, 1989.
Koumanova, R. et al., Efficient preservation of cosmetic emulsions and dental creams, in *Cosmetics and Toiletries Manufacturing Worldwide*, Aston Press, England, 1995.
Kundsin, R. B., Ed., *Airborne Contagion*, Annals NY Acad. Sciences, NY, 353, 1980.
Lachman, L., Urbanyl, T., and Weinstein, S., Stability of antibacterial preservatives in parenteral solutions. IV. Contribution of rubber closure composition on preservative loss, *J. Pharm. Sci.*, 52, 244, 1963.
Lamikanra, A. and Ogunbayo, T. A., A study of the antibacterial activity of butyl hydroxy anisole (BHA), *Cosmet. Toilet.*, 100, 69, 1985.
Lehne, R. K. (Moderator), Workshop—The preservation of cosmetic products, *CTFA Cosmet. J.*, 4, 2, 1972.
Levy, E., Insights into microbial adaptation to cosmetic and pharmaceutical products, *Cosmet. Toilet.*, 102, 69, 1987.
Lindstrom, S. M., Consumer use testing: Assurance of microbiological product safety, *Cosmet. Toilet.*, 101, 71, 1986.
Lindstrom, S. M. and Bowman, P. I., Resistance of *Pseudomonas* species to preservatives in cosmetic formulations, *J. Soc. Cosmet. Chem.*, 36, 5, 1985.
Lindstrom, S. M. and Hawthorne, J. D., Validating the microbiological integrity of cosmetic products through consumer-use testing, *J. Soc. Cosmet. Chem.*, 37, 481, 1986.
Lladres, C. S. and Ahearn, D. G., Synergistic inhibition of fungi by selected inhibitors, *J. Soc. Cosmet. Chem.*, 36, 5, 1985.
Lorenz, P., 5-bromo-5-nitro-1, 3-dioxane: A preservative for cosmetics, *Cosmet. Toilet.*, 92, 89, 1977.
Luck, E., Sorbic acid for the preservation of cosmetic preparations, *Soap Perfum. Cosmet.*, 981, 1964.
Luck, E. and Remmert, K., Sorbic acid and the preservation of cosmetic products, *Cosmet. Toilet.*, 107, 1993.
Lukacs, A. et al., Efficacy of a deodorant and its components: Triethyl citrate and perfume, *J. Soc. Cosmet. Chem.*, 42, 1991.
Madden, J. M. and Jackson, G. J., Cosmetic preservation and microbes: Viewpoint of the food and drug administration, *Cosmet. Toilet.*, 96, 75, 1981.
Maddox, D. N., Role of p-hydroxybenzoates in modern cosmetics, *Cosmet. Toilet.*, 97, 85, 1982.
Maddox, D. N., Development of New antimicrobial preservatives for toiletries, *Spec. Chem.*, 4, 487, 1988.
Maeda, Y. et al., High-performance liquid chromotographic determination of six p-hydroxybenzoic acid esters in cosmetics using sep-tak florisil cartridges for sample pre-treatment, *J. Chromatogr.*, 410, 413, 1987.
Maher, W. J. and Dietz, G. R., Radiation sterilization in the cosmetic industry, *Cosmet. Toilet.*, 96, 53, 1981.
Maibach, H. I. and Aly, R., Skin microbiology, NY:Springer-Verlag, 1981.
Manowitz, M., Preservation of cosmetic emulsions, *Sindar Reporter 1*, 1962.
Marchenay, Y. and Prevost, A., Application of pharmaceutical practices to the cosmetics industry, in *Cosmetics and Toiletries Manufacture Worldwide*, Aston Press, Hertfordshire, England, 1995.
Marinaro, A., Methods for evaluation of preservation systems in proprietary products, *Amer. Perf. Cosmet.*, 81, 29, 1966.

Marouchoc, S. R., Properties of Dowicil 200 preservative, *Cosmet. Toilet.*, 92, 91, 1977.
Marouchoc, S. R., Cosmetic preservation, *Cosmet. Tech.*, 38, 44, 1980.
Marples, R. R., Antibacterial cosmetics and the microflora of human skin, *Dev. Ind. Microbiol.*, 12, 178, 1971.
Maruzzella, J. C., Antifungal properties of perfume oils, *J. Amer. Pharm. Assn.*, 52, 601, 1963.
Marzulli, F. N., Evans, J. R., and Yoder, P. D., Induced *Pseudomonas* keratitis as related to cosmetics, *J. Soc. Cosmet. Chem.*, 23, 89, 1972.
Masse, M.O., Analysis of preservatives in and microbial cleanliness of baby products and products applied close to the eyes, *Cosmet. Toilet.*, 99, 46, 1984.
Matissek, R., Analysis of preservatives in cosmetics—methylisothiazolones, *Chromatographia*, 28, 34, 1989.
Matsuura, I., Methods for the prediction of shelf life, *Cosmet. Toilet.*, 96, 39, 1981.
McCarthy, T. J., Interaction between aqueous preservative solutions and their plastic containers, *Pharm. Weekbl.*, 105, 557, 1970a.
McCarthy, T. J., Interaction between aqueous preservative solutions and their plastic containers, *Pharm. Weekbl.*, 105, 1139, 1970b.
McCarthy, T. J., Microbiological control of cosmetic products, *Cosmet. Toilet.*, 95, 23, 1980.
McCarthy, T. J., Metal ions as microbial inhibitors, *Cosmet. Toilet.*, 100, 69, 1985.
McConville, J. F., Quaternium-15 as a cosmetic preservative, *Drug Cosmet. Ind.* 138, 52, 1986.
McConville, J. F. and Anderson, D. W., Aerobic microflora of the outer eye area of women in Los Angeles, CA, *J. Soc. Cosmet. Chem.*, 26, 83, 1975.
McConville, J. F. et al., Method For performing aerobic plate counts of anhydrous cosmetics utilizing Tween 60 and Arlacel 80 as dispersing agents, *Appl. Micro.*, 27, 5, 1974.
McIntyre, C. R., Hazard analysis critical control point (HACCP) identification, *Dairy Food, and Environmental Sanitation*, 1991.
McLaughlin, J. K., Zuckerman, B. D., Tenenbaum, S., and Wolf, B. A., Comparison of the API 20E, flow, and minitek systems for the identification of enteric and nonfermentative bacteria isolated from cosmetic raw materials, *J. Soc. Cosmet. Chem.*, 35, 253, 1984.
Meltzer, N. and Henkin, H., Gluteraldehyde—A preservative for cosmetics, *Cosmet. Toilet.*, 92, 1977.
Menne, T. and Hjorth, N., Routine patch testing with paraben esters, *Cont. Dermat.*, 19, 189, 1988.
Miller, W. S. and Korczynski, M. S., Microbiological controls in packaging, *Drug Cosmet. Ind.*, 110, 38, 1972.
Milstein, S. R., Orth, D. S., and Lichtin, J. L., Organic synthesis, antibacterial evaluation, and quantitative structure-activity relationships (QSAR) of cosmetic preservatives related to 5-bromo-5-nitro-1, 3-dioxane. I. Aliphatic analogs, *J. Soc. Cosmet. Chem.*, 35, 73, 1984.
Miyawaki, G. M., Patel, N. K., and Kostenbauder, H. B., Interaction of preservative with macromolecules, III. Parahydroxybenzoic acid esters in the presence of hydrophilic polymers, *J. Am. Pharm. Assoc, Sci. Ed.*, 48, 315, 1959.
Moberg, C. L. and Cohn, Z. A., Rene Jules Dubos, *Sci. Amer.*, 1991.
Moral, J., Cosmetic microbiology: New ingredients, new preservation strategies, *Cosmet. Toilet.*, 1992.

Morganti, P., Preservatives for cosmetic use: toxicity and sensitization problems, *J. Appl. Cosmetol.*, 6, 143, 1988.(Abstract)

Morse, L. J., Williams, H. L., Grenn Jr., F. P., Eldridge, E. E., and Rotta, J. R., Septicemia due to Klebsiella *pneumoniae* originating from hand cream dispenser, *N. Engl. J. Med.*, 277, 472, 1967.

Mulberry, G. K., Rapid screening methods for preservative efficacy evaluations, *Cosmet. Toilet.*, 102, 47, 1987.

Murray, E. A., A cosmetic preservative (Dowicil 200), *Det. Spec.*, 8, 31, 1971.

Muscatiello, M. J. and Penicnak, A. J., Historical perspective of cosmetic microbiology, *Cosmet. Toilet.*, 101, 47, 1986.

Muscatiello, M. J. and Penicnak, A. J., Evaluation of impedance microbiology, *Cosmet. Toilet.*, 102, 41, 1987.

Myers, G. E. and Pasutto, F. M., Microbial contamination of cosmetics and toiletries. *Can. J. Pharm. Sci.*, 8, 19, 1973.

Nelson, J. D. and Hyde, G. A., Sodium and zinc omadine antimicrobials as cosmetic preservatives, *Cosmet. Toilet.* 96, 87, 1981.

Neuwald, F. and Schmitzek, G., Storage studies on preservative solutions in glass and plastic bottles, *J. Mond. Pharm.*, 1, 11, 1968.

Nickerson, K., Kramer, V., and Kabara, J. J., The effectiveness of lauricidin preservative systems against detergent-resistant *Enterobacter cloacae*, *Soap Chem. Spec.*, Feb., 50, 1982.

Noble, W. C. and Savin, J. A., Steroid cream contaminated with *Pseudomonas aeruginosa*, *Lancet*, 1, 347, 1966.

O'Neill, J. J. and Mead, C. A., The parabens: Bacterial adaptation and preservative capacity, *J. Soc. Cosmet. Chem.*, 33, 75, 1982.

O'Neill, J. J., Peelor, P. L., Peterson, A. F., and Strube, C. H., Selection of parabens as preservatives for cosmetics and toiletries, *J. Soc. Cosmet. Chem.*, 30, 25, 1979.

Olson, J. C., Some considerations relative to microbial contamination of cosmetics, *Amer. Perf. Cosmet.*, 85, 43, 1970.

Olson, J. C., New horizons in microbiology, *FDA Papers*, 5, 17, 1971.

Olson, S. W., The application of microbiology to cosmetic testing, *J. Soc. Cosmet. Chem.*, 18, 191, 1967.

Orth, D. S., Linear regression method for rapid determination of cosmetic preservative efficacy, *J. Soc. Cosmet. Chem.*, 30, 321, 1979.

Orth, D. S., Establishing cosmetic preservative efficacy by use of D-values, *J. Soc. Cosmet. Chem.*, 31, 165, 1980.

Orth, D. S., Principles of preservative efficacy testing, *Cosmet. Toilet.*, 96, 43, 1981.

Orth, D. S., Preservative efficacy testing of cosmetic products: Rechallenge testing and reliablity of the linear regression method, *Cosmet. Toilet.*, 97, 61, 1982.

Orth, D. S., Adaptation of bacteria to cosmetic preservatives, *Cosmet. Toilet.*, 100, 57, 1985.

Orth, D. S., Determination of shampoo preservative stability and apparent activation energies by the linear regression method of preservative efficacy testing, *J. Soc. Cosmet. Chem.*, 38, 307, 1987.

Orth, D. S., Microbiological considerations in cosmetic formula development and evaluation, *J. Soc. Cosmet. Chem.*, 39, 5, 1988.

Orth, D. S., Effect of culture conditions and method of inoculum preparation on the kinetics of bacterial death during preservative efficacy testing, *J. Soc. Cosmet. Chem.*, 40, 193, 1989.

Orth, D. S., Synergism of preservative system components: Use of the survival curve slope method to demonstrate anti-*Pseudomonas* synergy of methyl paraben and acrylic acid homopolymer/copolymers *in vitro*, *J. Soc. Cosmet. Chem.*, 40, 347, 1989.

Orth, D. S., Microbiological considerations in cosmetic formula development and evaluation, *Cosmet. Toilet.*, 104, 49, 1989.

Orth, D. S., Rational development of preservative systems for cosmetic products, *Cosmet. Toilet.*, 104, 91, 1989.

Orth, D. S. and Milstein, S. R., Rational development of preservative systems for cosmetic products, *Cosmet. Toilet.*, 104, 91, 1989.

Owen, E. M., A method for the evaluation of preservative systems in cosmetic formulations, *TGA Cosmet. J.*, 1, 12, 1969.

Pader, M., *Oral Hygiene and Practice*, Marcel-Dekker, NY, 1987.

Pader, M., Toothpaste gels, *Cosmet. Toilet.*, 102, 1987.

Palmieri, M. J., FDA methodology for the microbiological analysis of cosmetics and topical drugs, *J. Soc. Cosmet. Chem.*, 34, 35, 1983.

Palmieri, M. J., Craito, S. L., and Meyer, R. F., Comparison of rapid NFT and API 20E with conventional methods for identification of gram-negative bacilli from pharmaceuticals and cosmetics, *Appl. Environ. Microbiol.*, 54, 838, 1988.

Parker, M. S., Some factors in the hygienic manufacture of cosmetics and their preservation, *Soap Perf. Cosmet.*, 43, 483, 1970.

Parker, M. S., The impact of microbiology on cosmetics, *Soap Perf. Cosmet.*, May, 303, 1970.

Parker, M. S., The microbiologically acceptable cosmetic, *Amer. Perf. Cosmet.*, 86, 35, 1971.

Parker, M. S., Ecology and preserved systems, *Cosmet. Toilet.*, 93, 49, 1978.

Parker, M. S., Microbial control of drugs and cosmetics, *Soap Perf. Cosmet.*, 44, 806, 1979.

Parker, M. S., Design and assessment of preservative systems for cosmetics, *Cosmet. Drug Preserv. Prin. Pract.*, 1, 389, 1984.

Parker, M. T., Clinical significance of the presence of microorganisms in pharmaceutical and cosmetic preparations (Review). *J. Soc. Cosmet. Chem.*, 23, 415, 1972.

Parsons, T., A microbiology primer for the microbiology manager, *Cosmet. Toilet.*, 105, 73, 1990.

Phillipson, I., Microbiology of cosmetics, *Mfg. Chem.*, 54, 45, 1983.

Phillipson, I. and McCulloch, T., EEC review, *Cosmet. Toilet.*, 101, 43, 1986.

Poltronieri, A., Privitera, S., and Salvi, A., Method for the determination of the microbiological assay of imiazolidinal urea, *J. Soc. Cosmet. Chem.*, 36, 313, 1985.

Pope, D. G. et al., Raw material microbial count reduction via cobalt-60 irradiation, *Pharm. Tech.*, Oct, 1978.

Preti, G. and Cutler, W. B., Human odors and their effects on the menstrual cycle, *J. Soc. Cosmet. Chem.*, 36, 5, 1985.

Proserpio, G., Protection of cosmetics by synergistic mixtures of preservatives, *Parf. Cosm. Sav. Fr.*, 2, 305, 1972.

Rastogi, S. C., Kathon CG content in cosmetic products, *Cont. Dermat.*, 19, 263, 1988.

Rastogi, S. C., Kathon CG and cosmetic products, *Cont. Dermat.*, 22, 155, 1990.

Reeve, P., Formulating techniques for the preservation with isothiazolones of cosmetics cotaining amines, *J. Appl. Cosmetol.*, 6, 149, 1988.
Rehn, D. et al., Cosmetic preservatives, *J. Soc. Cosmet. Chem.*, 31, 253, 1980.
Rheins, L. A. et al., Rapid induction of thy-1 antigenic markers on keratinocytes and epidermal immune cells in the skin of mice following topical treatment with common preservatives used in topical medications, *J. Invest. Dermatol.*, 89, 489, 1987.
Richard, G. et al., Qualitative analysis of preservatives using high-performance thin-layer chromatography, *Cosmet. Sci. Technol. Ser.*, 4, 157, 162, 1985.
Richardson, E. L., Update: Frequency of Preservative use in cosmetic formulas as disclosed to FDA, *Cosmet. Toilet.*, 96, 91,1981.
Rieger, M. M., Current aspects of cosmetic science: The inactivation of phenolic preservatives in emulsions, *Cosmet. Toilet.*, 96, 39, 1981.
Rieger, M. M., Role of water in performance of hydrophilic gums, *Cosmet. Toilet.*, 102, 1987.
Roderick, C. N., Gipson, M. V., and Kuhnley, L. C., Microbiological contamination of liquid bubble baths, *Cosmet. Toilet.*, 94, 99, 1979.
Rodgers, J. A. et al., Evaluation of methods for determining preservative efficacy, *CTFA Cosmet. J.*, 5, 2, 1973.
Rosen, W. E. and Berke, P. A., Modern concepts of cosmetic preservation, *J. Soc. Cosmet. Chem.*, 24, 663, 1973.
Rosen, W. E. and Berke, P. A., Germall 115 and nonionic emulsifiers, *Cosmet. Toilet.*, 94, 47, 1979.
Ryder, D. S., The thin layer chromatographic detection and determination of an imidazolidinyl urea antimicrobial preservative (Germall 115). *J. Soc. Cosmet. Chem.*, 25, 535, 1974.
Sabourin, J. R., Evaluation of preservatives for cosmetic products, *Drug Cosmet. Ind.*, Feb., 1991.
Sabourin, J. R., Selecting a preservative for creams and lotions, *Cosmet. Toilet.*, 101, 93, 1986.
Sakamoto, T. et al., Effects of some cosmetic pigments on the bactericidal activities of preservatives, *J. Soc. Cosmet. Chem.*, 38, 83, 1987.
Schanno, R. J. et al., Evaluation of 1, 3-dimethylol-5, 5-dimethyl hydantoin as cosmetic preservative, *J. Soc. Cosmet. Chem.*, 31, 85, 1980.
Schiller, I. et al., Microbial content of nonsterile therapeutic agents containing natural or seminatural active ingredients, *Appl. Micro.*, 16, 1924, 1968.
Schlitzer, R. L. and Rosenthal, R. A., Evaluating the antimicrobial efficacy of preservative systems: considerations for determining stability with regard to regulatory standards, *J. Soc. Cosmet. Chem.*, 31, 85, 1985.
Schmolka, I. R., Synergistic effects of nonionic surfactants upon cationic germicidal agents, *J. Soc. Cosmet. Chem.*, 24, 577, 1973.
Sciarra, J. J. and Cuties, A. J., Aerosols: The system of choice for cosmetics, *J. Soc. Cosmet. Chem.*, 39, 1988.
Scott, H., Methods for counting and testing for microorganisms in raw materials, topical and oral products, (Review), *J. Soc. Cosmet. Chem.*, 24, 65, 1973.
Shapiro, J. A., Multicellular behavior of bacteria, *ASM News*, March, 1991.
Shaw, A., Preservatives update, *Soap Cosmet. Chem. Spec.*, May, 1994.
Sherma, J. and Zorn, S., Determination of paraben preservatives in liquid products by quantitative preadsorbent thin layer chromatography with direct sample application, *Amer. Lab.*, 14, 26030, 1982.

Silber, P. M. et al., On the comedogenic potential of quaternium-15 and DMDM hydantoin, *J. Soc. Cosmet. Chem.*, 40, 135, 1989.

Singer, S., Use of preservative neutralizers in diluents and plating media, *Cosmet. Toilet.*, 102, 55, 1987.

Smart, R. and Spooner, D. F., Microbiological spoilage in pharmaceuticals and cosmetics, *J. Soc. Cosmet. Chem.*, 23, 721, 1972.

Smith, J. L., Product testing for preservation efficacy, *Cosmet. Toilet.*, 92, 30, 1977.

Smith, J. L., Preserving cosmetics, *Cosmet. Toilet.*, 96, 39, 1981.

Sonea, S., A bacterial way of life, *Nature* (Jan), 1988.

Sonea, S., A new look at bacteria, ASM News (Nov), 1989.

Sonea, S., Bacterial viruses, prophages, plasmids reconsidered, *Annals NY Acad. Sci.*, 503, 1987.

Sonea, S., The global organism, *The Sciences* NY Acad. Sci., Jul-Aug, 1988.

Spielmaker, R. J., Report of a cosmetic industry survey concerning correlation of preservative challenge and consumer use test results, *J. Soc. Cosmet. Chem.*, 36, 5, 1985.

Spooner, D. F., Problems in the preservation testing of topical products, *J. Appl. Cosmetol.*, 7, 93, 1989.

Stack, A. R. and Davis, H. M., Liquid chromatographic separation and fluorometric determination of quaternium-15 in cosmetics, *J. Assoc. Off. Anal. Chem.*, 67, 13, 1984.

Stehlin, D., Cosmetic allergies, *FDA Consum.*, 20, 28, 1986.

Steinberg, D. C., Botanical extracts and preservation issues, *Cosmet. Toilet.*, 106, 1991.

Steinberg, D. C., Cosmetic preservatives, *Drug Cosmet. Ind.*, 32, 109, 1984.

Steinberg, D. C., Preserving foundations, *Cosmet. Toilet.*, Feb., 1995.

Streek, M., Analysis of conservation agents (in cosmetics), *Parfeum. Kosmet.*, 71, 136, 1990.

Sykes, G. and Smart, R., Preservation of preparations for application to skin, *Amer. Perf. Cosmet.*, 84, 45, 1969.

Templars, C., The use of conductance microbiology in the cosmetics and toiletries industry, in *Cosmetics and Toiletris Manufacture Worldwide*, Aston Press, Hertfordshire, England, 1995.

Tenenbaum, S., Pseudomonads in cosmetics, *J. Soc. Cosmet. Chem.*, 18, 797, 1967.

Tenenbaum, S., Development and evaluation of an effective preservation system, *Proc. Joint Conf. Cosm. Sci. Toilet Goods Association*, Washington, D.C., 219, 1968.

Tenenbaum, S., Determination of the microbial content of water miscible cosmetic lotions: Preliminary report of a collaborative study, *TGA Cosmet. J.*, 2, 24, 29, 1970.

Tenenbaum, S., Significance of pseudomonads in cosmetic products, *Amer. Perf. Cosmet.*, 86, 33, 1971.

Tenenbaum, S., Microbial content of cosmetics and nonsterile drugs, *Cosmet. Perf.*, 88, 49, 1973.

Tenenbaum, S., Considerations leading to the development of the microbial guidelines of the CTFA., *Cosmet. Toilet.*, 92, 79, 1977.

Tenenbaum, S., Methodology for the national microbiological survey of cosmetics and toiletries, 1972-1975, *CTFA Cosmet. J.*, 9, 19, 1977.

Tenenbaum, S., Microbiological limit guidelines for cosmetics and toiletries, *CTFA Cosmet. J.*, 4, 25, 1991.

Thomas, M. J. and Majors, P. A., Animal, human, and microbiological safety testing of cosmetic products, *J. Soc. Cosmet. Chem.*, 24, 135, 1973.
Tosti, A. et al., Prevalence and sources of sensitization to emulsifiers: A clinical study, *Cont. Derm.*, 23, 68, 1990.
Totler, J. C., Preservative stability and preservative systems, *Int. J. Cosmet. Sci.*, 7, 157, 1985.
Tran, A. T., Hitchins, A. D., and Collier, S. W., Adequacy of cosmetic preservation: Chemical analysis, microbial challenge, and in-use testing, *Int. J. Cosmet. Sci.*, 16, 1994.
Tran, A. T., Hitchins, A. D., and Collier, S. W., Direct contact membrane method for evaluating preservative efficacy in solid cosmetics, *Int. J. Cosmet. Sci.*, 12, 175, 1990.
Troller, J. A., Methods to measure water activity, *J. Food Prot.*, Feb., 1983.
Troller, J. A., Water relations of food-borne bacterial pathogens - an updated review, *J. Food Prot.*, Aug., 1986.
Tuttle, E. et al, Preservation of protein solutions with 2-bromo-2-nitro-1, 3-propanediol (Bronopol), *Amer. Perf. Cosmet.*, 85, 87, 1970.
Van Abbe, J. V., Post-marketing surveillance to monitor the performance and safety-in-use of cosmetics and toiletries, *Cosmet. Toilet.*, 102, 61, 1987.
Van Abbe, N. J., Current problems in cosmetic science, *Pharm. J.*, 205, 669, 672, 1970.
Van Abbe, N. J., Preservatives and product safety, *Mfg. Chem.*, 58, 47, 1987.
Van Abbe, N. J. et al., The hygienic manufacture and preservation of toiletries and cosmetics, *J. Soc. Cosmet. Chem.*, 21, 719, 1970.
Van Doorne, H. and Claushuis, E. P. M., Enterobacteriaceae in pharmaceuticals (Holland), *Int. J. Pharm.*, 4, 119, 1979.
Vaughan, C. D., Solubility effects in product, package, penetration, and preservation, *Cosmet. Toilet.*, 103, 47, 1988.
Wachi, Y. et al., Decomposition of surface active agents by bacteria isolated from deionized water, *J. Soc. Cosmet. Chem.*, 31, 67, 1980.
Wallhaüsser, K. H., Sufactants and preservation of cosmetic preparations, *Surfactant Sci. Ser.*, 16, 211, 1985.
Wallhaüsser, K. H., The problem of preserving cosmetics., *Cosmet. Toilet.*, 91, 45, 1976.
Wallhaüsser, K. H., Microbiological quality control of skin care preparations, *Cosmet. Toilet.*, 93, 45, 1978.
Warwick, F. F., The role of preservatives in controlling microbial contamination introduced in the manufacture of cosmetics, *J. Appl. Cosmetol.*, Apr-Jun, 1993.
Weaver, J. E., Cardin, C. W., and Maibach, H. I., Dose-response assessments of Kathon biocide. I. Diagnostic use and diagnostic threshold patch testing with sensitized humans, *Cont. Dermat.*, 12, 141, 1985.
Wedderburn, D. L., Preservation of emulsions against microbial attack, *Adv. Pharm. Sci.*, 195, 268, 1964.
Wedderburn, D. L., Hygiene in manufacturing plant and its effect on the preservation of emulsions, *J. Soc. Cosmet. Chem.*, 16, 395, 1965.
Wilson, L. A. and Ahearn, D. G., *Pseudomonas*-induced corneal abscesses associated with contaminated eye mascaras, *Am. J. Ophthamol.*, 84, 112, 1977.
Wilson, L. A. et al., Microbial contamination in ocular cosmetics, *Am. J. Ophthamol.*, 67, 52, 1969.

Wilson, L. A., Kuehne, I. W., Hall, S. W., and Ahearn, L. G., Microbial contamination in ocular cosmetics, *Am. J. Ophthamol.*, 71, 1298, 1971.
Wolven, A. B. and Levenstein, I., Cosmetics—contaminated or not? *TGA Cosmet. J.*, 1, 34, 1969.
Wolven, A. B. and Levenstein, I., Microbiological examination of cosmetics, *Amer. Cosm. Perf.*, July, 1972.
Wooder, M. F., Introduction of a Cosmetic preservative: A safety perspective, *Seifen Oele, Fette Wachse*, 116, 368, 1990.
Woodford, and Adams, Ethanol and propylene glycol and mixtures of either with potassium sorbate vs. *P. aeruginosa* in o/w emulsions, *Amer. Cosm. Perf.*, 87, 53, 1972.
Woods, W. B., Nuosept C preservative for cosmetic and toiletry preparations, *Cosmet. Toilet.*, 100, 65, 1985.
Woodward, C. R., Some microbiological asoects of cosmetic manufacturing, *Amer. Perf. Cosmet.*, 86, 45, 1971.
Woodward, C. R. and McNamara, T. F., Microbiological consideration of cosmetic emulsions, *Amer. Perf. Cosmet.*, 85, 73, 1970.
Woodward, C. R. and McNamara, T. F., Practical method for the microbiological examination of cosmetics, *Amer. Perf. Cosmet.*, 86, 29,1971.
Wright, C. et al., Mutagenic activity of Kathon, an industrial biocide and cosmetics preservative containing 5-chloro-2-methyl-4-isothiazolin-3-one and 2-methyl-4-isothiazolin-3-one, *Mutation Research*, 119, 35, 1983.
Wyhowski de Bukanski, B. and Masse, M. O., Analysis of hexamidine, dibromopropamidine and chlorhexidine in cosmetic products by high-performance liquid chromatography, *Int. J. Cosmet. Sci.*, 6, 283, 1984.
Yablonski, J. I., Fundamental concepts of preservation, *Cosmet. Perf.*, 88, 39, 1973.
Yablonski, J. I., Strategies for cosmetic preservation, *Cosmet. Toilet.*, 92, 22, 1977.
Yablonski, J. I., Microbiological aspects of sanitary cosmetic manufacturing, *Cosmet. Toilet.*, 93, 37, 1978.
Yablonski, J. I. and Goldman, C. L., Microbiology of shampoos, *Cosmet. Perf.*, 90, 45, 1975.
Yamaguchi, M. et al., Antimicrobial activity of butylparaben in relation to its solubilization behavior by nonionic surfactants, *J. Soc. Cosmet. Chem.*, 33, 297, 1982.
Zeelie, J. J. and McCarthy, T. J., Antioxidants—multifunctional preservatives for cosmetic and toiletry formulations? *Cosmet.Toilet.*, 98, 51, 1983.
Zeelie, J. J. and McCarthy, T. J., Antioxidants—multifunctional preservatives for cosmetic and toiletry formulations? *Cosmet. Toilet.*, 98, 51, 1983.

APPENDIX 2

COMMONLY USED PRESERVATIVES IN COSMETIC MICROBIOLOGY

BENZALKONIUM CHLORIDE
EEC #: 54
CAS #: 139-07-1

Chemical names:	Mixture of alkyldimethylbenzylammonium chloride, N-dodecyl-N,N-dimethylbenzylammonium chloride
Trade Name(s):	Zephirol, Dodigen 226, Barquat MB-50.
Type of compound:	Quaternary, cationic

1. Structure and Chemical Properties

Appearance:	White or yellowish white amorphous powder
Odor:	Aromatic
Solubility:	Very soluble in water (about 50%) and alcohols
Optimum pH:	4.0 to 10.0; 1% solution gives a pH of 6.0 to 8.0
Stability:	Good, stable at 121°C for 30 min
Compatibility:	EDTA, nonionic detergents; incompatible with anionics, soap, nitrates, heavy metals, citrates, sodium tetraphosphate and hexametaphosphate, adsorbed by plastics
Neutralization:	Tween 80 and lecithin; reduced activity below pH of 5
Structural formula:	R is a mixture of the alkyls C_8H_{17} to $C_{18}H_{37}$

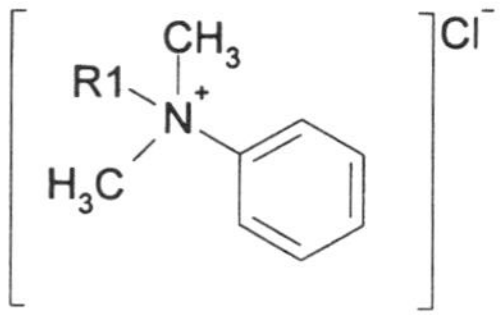

2. Antimicrobial Spectrum

Test organisms (10^6 CFU/ml)	Minimal germicidal concentration (μg/ml) (suspension test: for times shown)	
	24 hr	5 min
Staphylococcus aureus	4 to 10	50
Escherichia coli	10	80
Pseudomonas aeruginosa	10 to 100	200
Candida albicans	10	160
Aspergillus niger	100 to 200	—

3. Toxicity

Acute oral toxicity	Mouse:	LD_{50}, 300 mg/kg
	Rat:	LD_{50}, 450 to 750 mg/kg
Subacute oral toxicity	Rat:	Daily oral application of 550 ppm over 3 months without toxic effects
Chronic toxicity	Rat:	Dietary study with addition of 0.25% over 2 years without toxic effects
Primary skin irritation:	0.1% solution without effect on animals and humans; 0.5% caused irritation	
Rabbit eye irritation:	1:3000 dilution tolerated	
Mutagenicity:	Negative Ames test	

4. Cosmetic and Other Applications

In hair conditioners as a cationic, added as preservative, or in deodorants as antibacterial compounds. Antiseptic for preoperative skin preparation, wounds, burns; preservative in eyedrops.

5. Mode of Action

Solubilizes membrane lipids and other lipids by surfactant action to destroy cell integrity. May also denature proteins.

BENZETHONIUM CHLORIDE
EEC #: 53
CAS #: 121-54-0

Chemical names: N,N-Dimethyl-N-[2-[2-[4-(1,1,3,3-tetramethylbutyl)phenoxy]ethoxy]ethyl] benzene methane ammonium chloride; diisobutylphenoxyethoxyethy]dimethy] benzyl ammonium chloride
Trade Name(s): Hyamine 1622
Type of compound: Quaternary compound

1. Structure and Chemical Properties

Appearance: Thin, hexagonal crystals
Solubility: Very soluble in water, alcohol
Optimum pH: 4 to 10; 1% aqueous solution gives a pH of 4.8 to 5.5
Compatibility: Incompatible with soap, anionic detergents, mineral acids and salts
Neutralization: Tween 80 and lecithin
Structural formula:

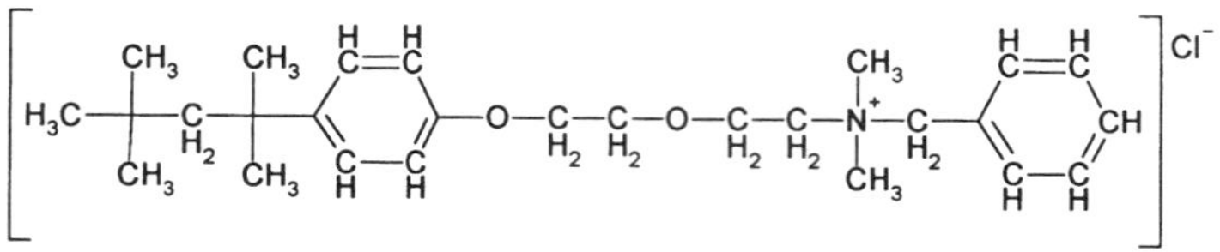

2. Antimicrobial Spectrum

Test Organisms (10^6 CFU/ml)	Minimal inhibitory concentration (μg/ml) (serial dilution test: 24-72 hr)	Minimal germicidal concentration (μg/ml) (suspension test: 10 min)
Staphylococcus aureus	40	0.5
Escherichia coli	50	32
Pseudomonas aeruginosa	800	250
Candida albicans	—	64
Aspergillus niger	—	128

3. Toxicity

Acute oral toxicity	Mouse:	LD_{50}, 500 mg/kg
	Rat:	LD_{50}, 420 mg/kg; 765 mg/kg
Human toxicity:	Ingestion may cause vomiting, collapse, convulsions, coma	

4. Cosmetic and Other Applications

In-use concentration (EEC), 0.1% as a preservative in deodorants Topical anti-infective, antiseptic, disinfectants, wound powders.

5. Mode of Action

Solubilizes membrane lipids and other lipids by surfactant action to destroy cell integrity. May also denature proteins.

BENZOIC ACID
EEC #: I/1
CAS #: 121-54-0

Chemical Names:	Benzene carboxylic acid
Trade Name(s):	
Type of compound:	Organic acid

1. Structure and Chemical Properties

Appearance:	Monoclinic tablets, plates, leaflets, or white powder
Solubility:	In water (20°C), 0.29%; in ethanol (20°C), 1 g in 2.3 ml; sodium salt in water (20°C), 1 g in 1.8 ml
Optimum pH:	2 to 5
Stability:	Stable at low pH
Compatibility:	Loss of activity in the presence of proteins and glycerol; incompatible with nonionics, quaternary compounds and gelatin
Neutralization:	pH above pKa
Structural formula:	

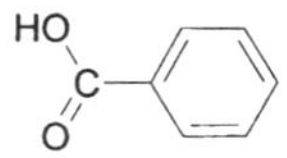

2. Antimicrobial Spectrum

Test organisms (10^6 CFU/ml)	Minimal inhibitory concentration (µg/ml) (serial dilution test: 24 to 72 hr; pH 6)	Minimal germicidal concentration (µg/ml) (suspension test: 24 to 72 hr; pH 6)
Staphylococcus aureus	50 to 100	20
Escherichia coli	100 to 200	160
Pseudomonas aeruginosa	100 to 200	160
Candida albicans	500 to 1000	1200
Aspergillus niger	500 to 1000	1000

The undissociated form has antimicrobial activity making it optimally effective below its pKa. In oil-water emulsions, benzoic acid migrates to the oil phase; only the amount dissolved in water is effective.

3. Toxicity

Acute oral toxicity	Mouse:	LD_{50}, 2.37 g/kg
	Rat:	LD_{50}, 1.7 g/kg
Subchronic toxicity	Mouse:	80 mg/kg per day over 3 months results in high mortality
Chronic toxicity:	40 mg/kg per day (mice and rats) up to 18 months inhibited growth	
Acceptable daily intake	Human:	0.5 mg/kg body weight per day
Human skin:	Toxic dose: 6 mg/kg	

4. Cosmetic and Other Applications

Use concentration, 0.1 to 0.2%; EEC maximum concentration, 0.5%. Preservative agent in food and pharmaceuticals (oral dosage forms).

5. Mode of Action

Destroys chemiosmotic balance across the cytoplasmic membrane by disruption of the membrane electrical potential through dissociation of protons from the compound into the cytoplasm of the cell. It may also denature proteins.

BENZYL ALCOHOL
EEC #: 51
CAS #: 100-51-6

Chemical names:	Benzyl alcohol, benzenemethanol, phenylcarbinol, phenylmelhanol
Trade Name(s):	
Type of compound:	Natural alcohol

1. Structure and Chemical Properties

Appearance:	Liquid
Odor:	Faint aromatic odor
Solubility:	4 g/100 ml water; 1 vol/1.5 vol 50% ethanol; miscible with absolute alcohol
Optimum pH:	Above 5
Stability:	Slowly oxidizes to benzaldehyde; dehydrates at low pH
Neutralization:	Inactivated by nonionics like Tween 80 and dilution
Structural formula:	

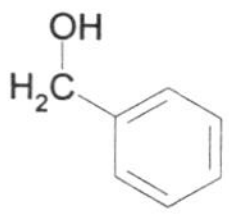

2. Antimicrobial Spectrum

Test organisms (10^6 CFU/ml)	Minimal germicidal concentration (μg/ml) (suspension test: 24 to 72 hr)
Staphylococcus aureus	25
Escherichia coil	2000
Pseudomonas aeruginosa	2000
Candida albicans	2500
Aspergillus niger	5000

3. Toxicity

Acute oral toxicity	Rat:	LD_{50}, 1.23 g/kg
	Mouse:	LD_{50}, 1.58 g/kg
	Rabbit:	LD_{50}, 1.94 g/kg
Acute dermal toxicity	Guinea pig:	LD_{50}, 5.0 ml/kg
	High percutaneous toxicity	
Human toxicity:	By dermal application; only a small amount resorbed by the dermis	
Toxicokinetic data:	Metabolized to hippuric acid	

4. Cosmetic and Other Applications

Use concentration, 1.0 to 3.0%; EEC guideline, 1,0%. Preservative in injectable drugs, ophthalmic products, and oral liquids.

5. Mode of Action

Disruption of membrane by solubilization of lipids and possibly denatures proteins.

5-BROMO-5-NITRO-L,3-DIOXANE
EEC #: 18
CAS #: 30007-47-7

Chemical name:	5-Bromo-5-nitro-l,3-dioxane
Trade Name(s):	Bronidox
Type of compound:	o-Acetal, o-formal

1. Structure and Chemical Properties

Appearance:	Solution, 10% (wt/vol) active ingredient in propylene glycol
Solubility (20°C):	25% in ethanol; 10% in isopropanol; 0.46% in water; 10% in propylene glycol
Optimum pH:	5 to 7
Stability:	Unstable at pH <5 and temperature above 50°C; corrosive to metal containers
Compatibility:	Compatible with nonionics
Neutralization:	Cysteine and protein;
Structural formula:	

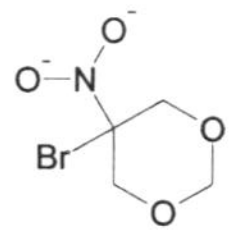

2. Antimicrobial Spectrum

Test organisms (10^6 CFU/ml)	Minimal inhibitory concentration (µg/ml) (serial dilution test: 24 to 72 hr)
Staphylococcus aureus	75
Escherichia coli	50
Pseudomonas aeruginosa	50
Candida albicans	25
Aspergillus niger	25

3. Toxicity

Acute oral toxicity	Mouse:	LD_{50}, 590 mg/kg
	Rat:	LD_{50}, 455 mg/kg
Subacute dermal tox.	Rat:	100 mg/kg-day no effect after 15 weeks; 200 mg/kg-day caused deaths
Primary skin irritation:	Irritation occurs above 0.5%	
Human skin:	Patch test at 0.1% without irritation; 0.5% in suspension and 0.25% in Vaseline showed irritation	
	Partial resorption (cutaneous), some metabolites in urine	
Rabbit eye irritation:	Ten applications of 0.05% produced no reaction; irritation threshold about 0.1%	
Guinea pig sens.:	No sensitization	

4. Cosmetic and Other Applications

In use concentration, 0.1% in EEC guideline; only for rinse-off products. Preservative for technical products

5. Mode of Action

Converts protein thiol groups to disulfides, resulting in denaturation of proteins with sulfhydryl groups at the active sites.

2-BROMO-2-NITROPROPANE-1,3-DIOL
EEC #: 19
CAS #: 52-51-7

Chemical name:	2-Bromo-2-nitropropane-1,3-diol
Trade Name(s):	Bronopol
Type of compound:	Alcohol

1. Structure and Chemical Properties

Appearance:	White crystalline powder
Odor:	Faint characteristic odor
Solubility:	Water, 25% (wt/vol); ethanol, 50%; isopropanol, 25%; glycerol, 1%; propyleneglycol, 14%
Optimum pH:	5.0 to 7.0
Stability:	Stable at low pH; yellows and browns under alkaline conditions; unstable with iron and aluminum; stable with stainless steel and tin; nitrite evolved to form nitrosamine
Compatibility:	Not affected by anionic, cationic, nonionic surfactants, or proteins
Neutralization:	0.1% cysteine; thioglycolate and thiosulfate also inactivate
Structural formula:	

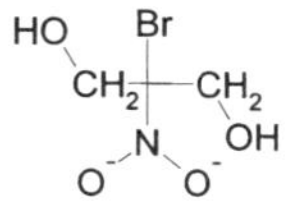

2. Antimicrobial Spectrum

Test organisms (10^6 CFU/ml)	Minimal inhibitory concentration (μg/ml) (serial dilution test; 24 to 72 hr incubation time)
Staphylococcus aureus	62.5
Escherichia coli	31.25
Pseudomonas aeruginosa	31.25
Candida albicans	50
Aspergillus niger	50

3. Toxicity

Acute oral toxicity	Mouse (male): LD_{50}, 374 mg/kg
	Rat (male): LD_{50}, 307 mg/kg
Intraperitoneal tox.	Rat (male): LD_{50}, 22 mg/kg
Acute dermal toxicity:	Rat (acetone solution): death at 160 mg/kg
Primary skin irritation:	0.5% in acetone, 2.5% in aq. methylcellulose, and 5.0% in polyethoxyleneglycol nonirritant
Rabbit eye irritation:	0.5% in saline or 2% in polyethyleneglycol nonirritant 5% is irritant
Guinea pig sens.	Three challenges for 2/10 animals sensitized. (Magnusson-Kligman)
Human sensitization:	Human skin irritant at 0.25% to 1% in soft paraffin and at 0.25% in aqueous buffer at pH 5.5
Chronic toxicity:	In 90-day test, daily oral doses of 20 mg/kg to rats were tolerated; 80 to 160 mg/kg caused gastrointestinal lesions, respiratory distress, and some deaths
Carcinogenicity:	0.2 to 0.5% in 0.3 ml acetone applied to shaved backs of mice 3x/week for 80 weeks did not increase spontaneous tumor profile. 10 to 160 mg/kg-day orally for 2 yr without tumor incidence.
Mutagenicity:	No mutagenic activity (Ames or host-mediated assay in mice)

4. Cosmetic and Other Applications

Used at 0.01 to 0.1% in hand and face creams, shampoos, hair dressings, mascaras, and bath oils, pharmaceutical products, household products (fabric conditioners and washing detergents).

5. Mode of Action

Forms disulfide bonds with thiol groups to denature proteins. For example, it inhibits dehydrogenase activity as a result.

CHLOROBUTANOL
EEC #: I/11
CAS #: 57-15-8

Chemical name:	1,1,1-trichloro-2-methylpropan-2-ol
Trade Name(s):	
Type of compound:	Chlorinated alcohol

1. Structure and Chemical Properties

Appearance:	Colorless crystals
Odor:	Camphor-like odor
Solubility:	0.8% in water; 1 g/ml ethanol or propylene glycol; 1 g/10 ml glycerin
Optimum pH:	Acidic pH (up to 4.0)
Stability:	Decomposed by alkali and heat
Compatibility:	Incompatible with nonionics, alkali; unstable in polyethylene
Neutralization:	Tween 80 and polyvinylpyrrolidone
Structural formula:	

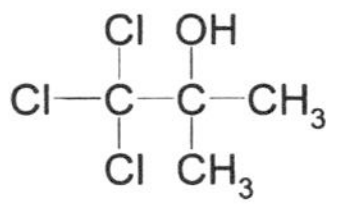

2. Antimicrobial Spectrum

Test organisms (10^6 CFU/ml)	Minimal germicidal concentration (μg/ml) (suspension test: 24 to 72 hr)
Staphylococcus aureus	625
Escherichia coli	625
Pseudomonas aeruginosa	1000
Candida albicans	2500
Aspergillus niger	5000

3. Toxicity

Acute oral toxicity Dog: LD_{50}, 238 mg/kg

4. Cosmetic and Other Applications

Use concentration up to 0.5%; prohibited in aerosols except for foams. Label must state "contains chlorobutanol."

5. Mode of Action

Disruption of membrane by alcohol solubilization of lipids and possibly denatures proteins.

p-CHLORO-m-CRESOL
EEC #: 26
CAS #: 59-50-7

Chemical Name:	4-chloro-3 methylphenol
Trade Name(s):	Preventol CMK
Type of compound:	Halogenated phenolic

1. Structure and Chemical Properties

Appearance:	White powder
Odor:	Odorless when very pure
Solubility:	1 g/260 ml of water at 20°C; freely soluble in alcohol
Optimum pH:	Broad-spectrum activity at acidic pH
Stability:	Aqueous solution yellows in light and air
Compatibility:	Partial inactivation in the presence of nonionics; discoloration with iron salts
Neutralization:	Dilution
Structural formula:	

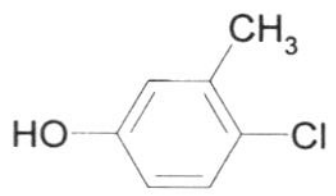

2. Antimicrobial Spectrum

Test organisms (10^6 CFU/ml)	Minimal inhibitory concentration (μg/ml) (serial dilution test; 24 to 72 hr incubation time)
Staphylococcus aureus	625
Escherichia coli	1250
Pseudomonas aeruginosa	1250
Candida albicans	2500
Aspergillus niger	2500

3. Toxicity

Acute oral toxicity	Mouse: LD_{50}, 4 g/kg
Guinea pig sens:	None

4. Cosmetic and Other Applications

Use concentration, 0.1 to 0.2% in protein shampoos and baby cosmetics; EEC maximum, 0.2%. Topical antiseptic, disinfectant, preservative in pharmaceutical products.

5. Mode of Action

As with many chlorinated phenolics PCMC most likely uncouples oxidation from phosphorylation and inhibits active transport by disrupting the cell membrane through solubilizing lipids and denaturing proteins. Once membrane integrity is compromised, the cell is more permeable to protons and thus any potential gradient for running ATP synthetase is destroyed.

CHLORHEXIDINE DIGLUCONATE
EEC #: 31
CAS #: 55-56-7

Chemical name:	Bis(p-chlorophenyldiguanido)hexane
Trade Name(s):	Hibitane, Novalsan, Rotasept, Sterilon, Hibiscrub, Arlacide
Type of compound:	Biguanide

1. Structure and Chemical Properties

Appearance:	White crystalline powder
Odor:	Odorless
Solubility:	Digluconate in HOH, >70%; diacetate and dihydrochloride salts are less soluble
Optimum pH:	5 to 8
Stability:	Unstable above 70°C
Compatibility:	Compatible with cationics; incompatible with anionics, gums, soaps, inorganic anions, alginates, carboxymethylcellulose, cork seals
Neutralization:	Nonionic surfactants
Structural formula:	

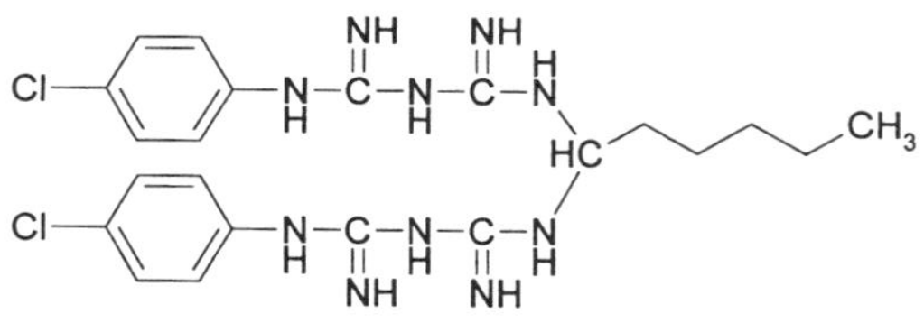

2. Antimicrobial Spectrum

Test organisms (10^6 CFU/ml)	Minimal inhibitory concentration (μg/ml) (serial dilution test: 24 to 72 hr)	Minimal germicidal concentration (μg/ml) (suspension test; 24 to 72 hr)
Staphylococcus aureus	0.5 to 1.0	100
Escherichia coli	1.0	100
Pseudomonas aeruginosa	5 to 60	400
Candida albicans	10 to 20	400
Aspergillus niger	200	400

3. Toxicity

Acute oral toxicity	Mouse: LD_{50}, 2 g/kg (diacetate)
Chronic toxicity	Rat: 0.05% in drinking water over 2 years without toxic effects
Primary skin irritation:	No effect
Human sensitization:	Allergic tendency
Mutagenicity:	Positive Ames test; positive DNA repair test

4. Cosmetic and Other Applications

Use concentrations 0.01 to 0.1%. Used in nonionic creams, toothpaste, deodorants/antiperspirants; EEC max., 0.3%. Skin disinfectant, preservative in eye-care products at 0.01%; topical antiseptic at 0.02%.

5. Mode of Action

General protein coagulant/denaturant that destroys membrane integrity and cytoplasmic enzymes. The molecule may become oriented within a lipid component of the membrane to cause a general disruption of membrane structure and function.

CHLOROXYLENOL
EEC #: 32
CAS #: 88-04-0

Chemical Name:	p-chloro-meta-xylenol, PCMX
Trade Name(s):	Ottasept; Nipacide
Type of compound:	Halogenated phenolic

1. Structure and Chemical Properties

Appearance:	Crystalline powder
Odor:	Phenolic
Solubility:	0.33 g in 1 liter HOH at 20°C
Optimum pH:	Wide
Compatibility:	Incompatible with cationics and non ionics
Neutralization:	Dilution
Structural formula:	

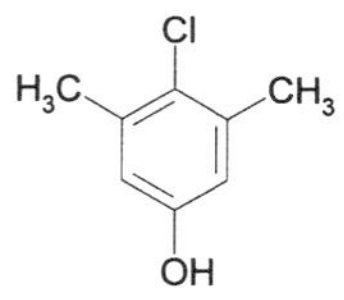

2. Antimicrobial Spectrum

Test organisms (10^6 CFU/ml)	Minimal inhibitory concentration (µg/ml) (serial dilution test: 24 to 72 hr)
Staphylococcus aureus	250
Escherichia coli	1000
Pseudomonas aeruginosa	1000
Candida albicans	2000
Aspergillus niger	2000

3. Toxicity

Primary skin irritation:	Less than phenol or cresol
Guinea pig sens.:	No sensitization

4. Cosmetic and Other Applications

Used for protein solutions, hair conditioners, silicone emulsions. EEC, 0.5%. Maximum concentration in soap, 2%. Used as an antiseptic, as a preservative for pharmaceutical products, and as an ingredient in disinfectants.

5. Mode of Action

As with many chlorinated phenolics chloroxylenol most likely uncouples oxidation from phosphorylation and inhibits active transport by disrupting the cell membrane through solubilizing lipids and denaturing proteins. Once membrane integrity is compromised, the cell is more permeable to protons and thus any potential gradient for running ATP synthetase is destroyed.

POLYAMINOPROPYL BIGUANIDE
EEC #: 42

Chemical Name:	polyhexamethylene biguanide hydrochloride
Trade Name(s):	Cosmocil CQ (ICI Americas, Inc.)
Type of compound:	Cationic

1. Structure and Chemical Properties

Appearance:	Clear, yellowish liquid
Odor:	Odorless
Solubility:	Water and alcohol soluble
Optimum pH:	4 to 8
Stability:	Stable below 80°C
Compatibility:	See chlorhexidine data
Neutralization:	See chlorhexidine data
Structural formula:	

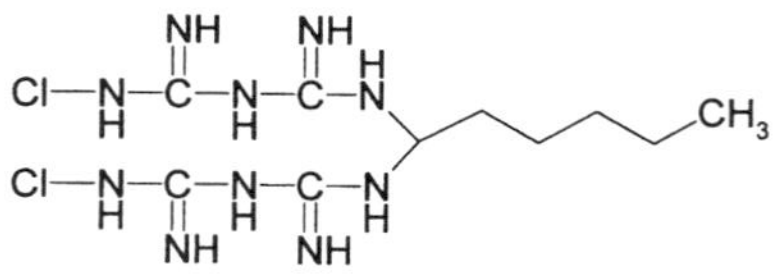

2. Antimicrobial Spectrum

Test organisms (10^6 CFU/ml)	Minimal inhibitory concentration (μg/ml) (serial dilution test: 24 to 72 hr)
Staphylococcus aureus	20
Escherichia coli	20
Pseudomonas aeruginosa	100
Candida albicans	—
Aspergillus niger	375

3. Toxicity

Acute oral toxicity	Rat: LD_{50}, 5 g/kg
Chronic toxicity:	90-day feeding test, 3.1 ppm for 90 days showed no toxicity or abnormalities. At 6.2 ppm, rats showed retardation of growth and lower food intake. Two and one-half year feeding test with pathogen-free Wistar rats at dietary levels of 3000, 5000, and 10,000 ppm. During the first 3 months, reduction in body weight and food intake; no-effect level, 5000 ppm.
Primary skin irritation:	Concentrated form is a strong irritant. 50,000 ppm tolerated by rats with no irritation. For mice the no-effect level was 100 mg/kg per day.
Rabbit eye irritation:	No irritant effect with 0.1 ml of a 2000 ppm dilution.
Photoirritation:	No significant photoirritancy.
Guinea pig sens.:	No sensitization.
Environmental tox.:	Ralnbow trout: 96h LC_{50}, 10 ppm

4. Cosmetic and Other Applications

Use concentration, 0.2 to 1.0% (of 20% solution): EEC, 0.3%. Used as disinfectant and preservative for technical products.

5. Mode of Action

See chlorhexidine data

DEHYDROACETIC ACID
EEC #: 4
CAS #.: 520-45-6

Chemical Name(s):	3-acetyl-6-methyl-2H-pyran-2,4(3H)dione
Trade Name(s):	
Type of compound:	Organic Acid

1. Structure and Chemical Properties

Appearance:	Sodium salt is a colorless, tasteless powder
Odor:	Odorless
Solubility:	The acid form is relatively insoluble; the salt (wt/wt at 25°C)is 1% in ethanol, 48% in propylene glycol, and 33% in water
Optimum pH:	5-6.5. Activity decreases with higher pH above the pKa
Stability:	Stable to heat.
Neutralization:	pH above 6.5.
Structural formula:	

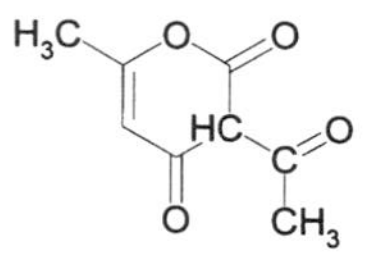

2. Antimicrobial Spectrum

Test organisms (10^6 CFU/ml)	Minimal inhibitory concentration (µg/ml) (serial dilution test; 24 to 72 hr)	Minimal germicidal concentration (µg/ml) (suspension test: 24 to 72 hr)
Staphylococcus aureus	10,000	20,000
Escherichia coli	10,000	20,000
Pseudomonas aeruginosa	>20,000	20,000
Candida albicans	200	ND
Aspergillus niger	200	ND

Only undissociated dehydroacetic acid is active. This activity depends on pH being below the pKa. Data above is for pH=6

3. Toxicity

Acute oral toxicity	Rat:	LD_{50}, 1.0 g/kg (sodium salt)
	Human:	Impaired kidney function; vomiting, ataxia, and convulsions.
Chronic toxicity:	Rat:	Daily dose with food, 0.1%, over 2 years showed no toxic effect; no-effect level, >50 mg/kg
Primary skin irritation:	No irritation or sensitization in humans.	

4. Cosmetic and Other Applications

0.02 to 0.2%. EEC maximum, 0.6%. Preservative for pumpkins in the US. Not allowed in Europe for food preservation.

5. Mode of Action

As with most organic acids they destroy the chemiosmotic balance across the cytoplasmic membrane by disruption of the membrane electrical potential through dissociation of protons from the compound into the cytoplasm of the cell. DHA may also denature proteins.

DICHLOROBENZYL ALCOHOL
EEC #: 24
CAS #.: 1777-82-8

Chemical Name(s):	2, 4-dichlorobenzyl alcohol
Trade Name(s):	Myacide SP, Unikon A-22
Type of compound:	Chlorinated Alcohol

1. Structure and Chemical Properties

Appearance:	White to yellowish crystalline powder
Solubility:	Water (20°C)-0.1%, propylene glycol-73.0%; soluble in ethanol and isopropanol.
Optimum pH:	Wide pH range (3.0-9.0)
Stability:	Can oxidize in aqueous solutions
Compatibility:	Incompatible with some anionics and nonionics
Neutralization:	Dilution
Structural formula:	

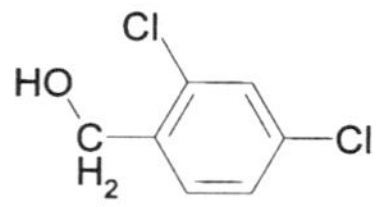

2. Antimicrobial Spectrum

Test organisms (10^6 CFU/ml)	Minimal inhibitory concentration (μg/ml) (serial dilution test: 24 to 72 hr)
Staphylococcus aureus	1000
Escherichia coli	500
Pseudomonas aeruginosa	1000
Candida albicans	500
Aspergillus niger	500

3. Toxicity

Acute oral toxicity	Mouse:	LD_{50}, 2.3 g/kg
	Rat:	LD_{50}, 3.0 g/kg
Subchronic toxicity:	Rat:	98-day test, 7.2 and 14.4 ppm daily with the diet—no toxic effect
Primary skin irritation:	Rabbit:	0.5% solution to shaved flanks of rabbits for 5 days was not irritating
Rabbit eye irritation:	0.8% in aqueous solution had no effect	
Guinea pig sens.:	None	
Mutagenicity:	Negative Ames test	
Environmental tox.:	*Daphnia magna*:	LC_{50} (24 hr), 22.0 ppm; (48 hr), 13.1 ppm
	Rainbow trout:	LC_{50}, (24 hr), 18.9 ppm; (48 hr), 14.4 ppm; (72 hr), 13.3 ppm
	Mallard duck:	Acute oral LD_{50}, 2.5 mg/kg

4. Cosmetic and Other Applications

Use concentration 0.15% in aqueous solutions, lotions, creams and gel formulations. Topical antiseptic, disinfectant, antiseptic mouthwash and gargle.

5. Mode of Action

Disruption of membrane by alcohol solubilization of lipids and possibly denatures proteins.

DMDM HYDANTOIN
EEC #: 50
CAS #: 6440-58-0

Chemical Name(s):	Dimethyloldimethylhydantoin
Trade Name(Supplier):	Glydant, Dekafald, Mackstat DM, Nipaguard DMDMH
Type of compound:	Formaldehyde donor

1. Structure and Chemical Properties

Appearance:	Clear solution
Odor:	Mild formaldehyde odor
Solubility:	Freely soluble in water (>50%) and ethanol
Optimum pH:	3.5 to 10.0
Stability:	Stable over wide pH range and temperature conditions (<90°C)
Compatibility:	Compatible with anionics, cationics, nonionics, and proteins
Neutralization:	Dilution and peptone
Structural formula:	

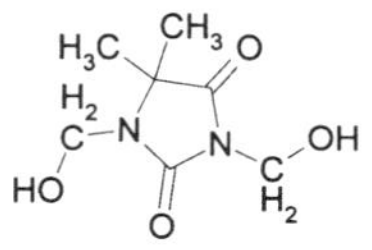

2. Antimicrobial Spectrum

Test organisms (10^6 CFU/ml)	Minimal inhibitory concentration (μg/ml) (serial dilution test: 24 to 72 hr)	Minimal germicidal concentration (μg/ml) (suspension test: 24 to 72 hr)
Staphylococcus aureus	250 to 800	4000
Escherichia coli	500	4000
Pseudomonas aeruginosa	800 to 1000	4000
Candida albicans	725 to 1250	5000
Aspergillus niger	750 to 1500	ND

Broad spectrum mainly against bacteria. Should be combined with antifungal components (e.g., anionic surfactants, parabens, Kathon CG, or formaldehyde) for products needing protection from yeasts and molds.

3. Toxicity

Acute oral toxicity	Rat (female):	LD_{50}, 3.8 g/kg
	Rat (male):	LD_{50}, 2.7 g/kg
Acute dermal toxicity:	Rabbit:	LD_{50}, >20 g/kg
Primary skin irritation:	No effect (rabbit)	
Human patch test:	400 ppm in water over 9 to 24 h occluded patch showed no irritation	
Rabbit eye irritation:	No irritation using 1.0% wt/vol solution	
Phototoxicity:	Not a photoallergen	
Sensitization:	None with 4000 ppm on 50 people	
Mutagenicity:	Ames test shows nonmutagenic	

4. Cosmetic and Other Applications

Use concentration, 0.15 to 0.4% shampoos, conditioners, hand creams. Preservative agent in detergents.

5. Mode of Action

Mechanism of action is the same as for formaldehyde which reacts with proteins in the membrane and cytoplasm to denature them

ETHANOL OR ALCOHOL
CAS #: 64-17-5

Chemical Name(s):	Ethanol refers to absolute ethyl alcohol. Alcohol refers to 95% (vol/vol) ethanol.
Trade Name(s):	
Type of compound:	Alcohol

1. Structure and Chemical Properties

Appearance:	Clear, colorless.
Odor:	Characteristic, burning taste.
Solubility:	Miscible with water, acetone, and glycerol.
Optimum pH:	Acidic
Stability:	Absorbs water; volatile
Compatibility:	Incompatible with acacia, albumin, bromine, chlorine;
Neutralization:	Inactivated by dilution to below 1%; may be inactivated by nonionics
Structural formula:	

$H_3C{-}\underset{H_2}{C}{-}OH$

2. Antimicrobial Spectrum

Test organisms (10^6 CFU/ml)	Minimal inhibitory concentration (%) (serial dilution test; 24 to 72 hr)	Minimal killing time (sec) 70% concentration (suspension test)
Staphylococcus aureus	5%	15
Escherichia coli	5%	30
Pseudomonas aeruginosa	5%	10
Candida albicans	5%	ND
Aspergillus niger	5%	ND

3. Toxicity

Acute oral toxicity	Rat:	LD_{50},13.7 g/kg
	Guinea pig:	LD_{50},5.5 g/kg
	Rabbit:	LD_{50},9.5 g/kg
Chronic toxicity:	Daily tolerable dose, 80 g. Acceptable dally intake, 7 g/kg per day	
Primary skin irritation:	Predictive skin sensitization test showed delayed allergic skin reaction with 50% solution	

4. Cosmetic and Other Applications

Satisfactory preservation above 15%, better at 20%. Use at 60 to 70% for disinfection. Topical anti-infective, antiseptic.

5. Mode of Action

Disruption of membrane by alcohol solubilization of lipids and possibly denatures proteins.

FORMALDEHYDE
EEC #: 1/5
CAS #.: 50-00-0

Chemical Name(s):	Formaldehyde, Formalin
Trade Name(s):	
Type of compound:	Aldehyde, 37% (wt/wt) in water; formalin has 10% methanol added to prevent polymerization

1. Structure and Chemical Properties

Appearance:	Colorless liquid; powerful reducing agent especially in presence of alkali; keep in well closed container; density, 1.12 g/ml.
Odor:	Pungent
Solubility:	Freely soluble in water
Optimum pH:	pH 3 to 10; Formalin pH 2.5 to 4.0
Stability:	May become cloudy in cold and form trioxymethylene (paraformaldehyde);oxidizes to formic acid.
Compatibility:	Incompatible with ammonium, hydrogen peroxide, iodine, iron, gelatin, proteins
Neutralization:	Dilution, peptone, and ammonium ions
Structural formula:	

$H_2C{=}O$

2. Antimicrobial Spectrum

Test organisms (10^6 CFU/ml)	Minimal inhibitory concentration (μg/ml) (serial dilution test: 24 to 72)	Minimal germicidal concentration (μg/ml) (suspension test: 24 to 72)
Staphylococcus aureus	125	62.5
Escherichia coli	125	31.25
Pseudomonas aeruginosa	125	62.5
Candida albicans	500	250
Aspergillus niger	500	500

3. Toxicity

Acute oral toxicity	Rat:	LD_{50}, 800 mg/kg	Guinea pig:	LD_{50}, 260 mg/kg
Acute subcutaneous toxicity:	Mouse:	LD_{50}, 300 mg/kg		
	Dog:	LD_{50}, 800 mg/kg		
Inhalation toxicology:	Rat:	LD, 250 mg/kg (4 hr)		
	Man:	LD, 36 mg/kg		

Human toxicity: Current permissible exposure limit: 3 ppm (OSHA); Irritation (eye, nose, throat): 0.03 to 4.0 ppm; allergy caused by inhalation exposure to 0.3% solutions; 1 to 4% people show positive patch test with 2% formaldehyde; 44% people sensitized show positive patch test to 30 ppm solution .

Mutagenicity: Ames test negative; mouse lymphoma positive; sister chromatid exchange positive. Chromosomal aberrations in bone marrow negative; *Drosophila* positive

Carcinogenicity: Rat: inhalation of 2 ppm formaldehyde 6 hr/day, 5 days/week for 24 months no effect; at 6-15 ppm, 1.5% to 43.2% tumor frequency

4. Cosmetic and Other Applications

Preservative for shampoos (0.1 to 0.2%). Label warning: "Contains Formaldehyde" when > 0.05% (EEC).

5. Mode of Action

Denatures proteins by reacting with amino groups in proteins of the cell wall, membrane, and cytoplasm.

DIAZOLIDLNYL UREA
CAS #: 78491-02-8

Chemical Name(s):	N-(Hvdroxymethyl)-N-(1,3-dihydroxymethyl-2,5-dloxo-4-imidazolidinyl)-N'-(hydroxymethyl) urea
Trade Name(s):	Germall II
Type of compound:	Heterocyclic imidazolidinyl urea, formaldehyde donor

1. Structure and Chemical Properties

Appearance:	Fine. white, free-flowing powder
Odor:	None or characteristically mild
Solubility:	Water soluble
Optimum pH:	Wide range
Stability:	Stable
Compatibility:	Compatible with ionics and non-ionics, protein
Neutralization:	Dilution and peptone
Structural formula:	

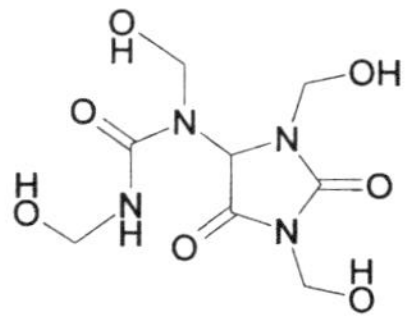

2. Antimicrobial Spectrum

Test organisms (10^6 CFU/ml)	Minimal inhibitory concentration (µg/ml) (serial dilution test; 24 to 72 hr)	Minimal germicidal concentration (µg/ml) (suspension test: 24 to 72 hr)
Staphylococcus aureus	250	1000
Escherichia coli	1000	4000
Pseudomonas aeruginosa	1000	4000
Candida albicans	8000	8000
Aspergillus niger	4000	8000

3. Toxicity

Acute oral toxicity	Rat:	LD_{50}, 2.57 g/kg
Acute dermal toxicity	Rabbit:	LD_{50}, >2.0 g/kg
Primary skin irritation:	Rabbit:	1 or 5% solution not an irritant
Rabbit eye irritation:	1 or 5% solution not an irritant	
Guinea pig sens.:	No sensitization	

4. Cosmetic and Other Applications

Use at 0.1 to 0.5% in combination with parabens or other antifungals

5. Mode of Action

As the chemical degrades, it donates formaldehyde to the microorganism. The formaldehyde denatures proteins by reacting with amino groups in proteins of the cell wall, membrane, and cytoplasm.

GLUTARALDEHYDE

Chemical Name(s):	Glutaraldehyde
Trade Name(Supplier):	Ucaricide
Type of compound:	Dialdehyde

1. Structure and Chemical Properties

Appearance:	Oily liquid (25% solution) stabilized with ethanol, alkaline pH, or 0.1 to 0.25% hydroquinone
Odor:	Pungent odor
Solubility:	Slightly soluble in water
Optimum pH:	Broad pH range: optimal pH for bactericidal activity 7.5 to 8.5
Compatibility:	Inactivated by ammonia or primary amines at neutral to basic pH
Neutralization:	Sodium bisulfite
Structural formula:	

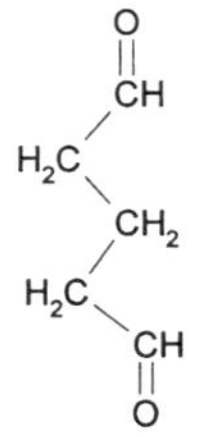

2. Antimicrobial Spectrum

Compared with 4% aqueous formaldehyde, 2% aqueous glutaraldehyde is 10 times as effective as a bactericidal and sporicidal agent.

3. Toxicity

Inhalation toxicity	Rat:	8 hr in saturated glutaraldehyde atmosphere caused no deaths.
Acute oral toxicity	Rat:	LD_{50}, 60 mg/kg
Chronic toxicity:		Not mutagenic
Primary skin irritation:		LD_{50} (25% solution), 2.38 mg/kg
Human contact dermatitis:		550 ppm without effect (706 persons)

4. Cosmetic and Other Applications

Use concentration, 0.02 to 0.2%. Disinfectant for instruments and equipment; preservative at 0.2% and pH about 5; corrosive.

5. Mode of Action

Denatures proteins by reacting with amino groups in proteins of the cell wall, membrane, and cytoplasm.

HEXACHLOROPHENE
EEC #: I/6
CAS #.:70-30-4

Chemical Name(s):	2,2′-Dihydroxy-3,3′,5,5′,6,6′-hexachlorodiphenylmethane, 2,2′-methylenebis[3,4,6-trichlorophenol]
Trade Name(s):	
Type of compound:	Chlorinated phenolic

1. Structure and Chemical Properties

Appearance:	White crystals
Odor:	None
Solubility:	Practically insoluble in water; soluble in alcohol, propylene glycol, polyethylene glycols
Optimum pH:	pH 5-6 for bacteriostatic effect; for bacteriostasis pH 8
Compatibility:	Incompatible with Tween
Neutralization:	Tween 80
Structural formula:	

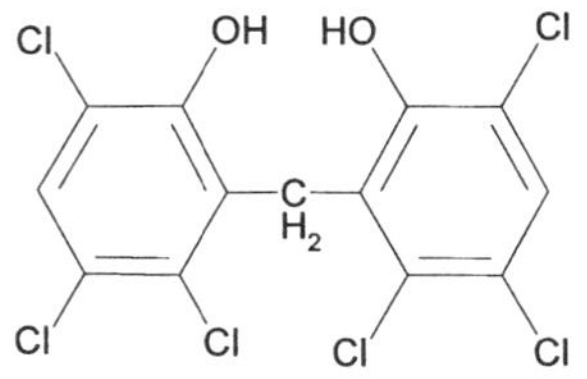

2. Antimicrobial Spectrum

Test organisms (10^6 CFU/ml)	Minimal inhibitory concentration (μg/ml) (serial dilution test; 24 to 72 hr)
Staphylococcus aureus	—
Escherichia coli	12.5
Pseudomonas aeruginosa	250
Candida albicans	1000
Aspergillus niger	300

3. Toxicity

Chronic toxicity: Potential neurotoxicity in humans, the FDA has regulated its use

4. Cosmetic and Other Applications

EEC guideline maximum dose for cosmetics, 0.1%; not to be used in preparations for children and intimate hygiene; warnings on the label: Not To Be Used for Babies. Contains Hexachlorophene; not for children under 3 years of age

5. Mode of Action

As with many chlorinated phenolics hexachlorophene most likely uncouples oxidation from phosphorylation and inhibits active transport by disrupting the cell membrane through solubilizing lipids and denaturing proteins. Once membrane integrity is compromised, the cell is more permeable to protons and thus any potential gradient for running ATP synthetase is destroyed.

HEXAMETHYLENETETRAMINE
EEC #: 44
CAS #.100-97-0

Chemical Name(s):	1,3,5,7-Tetraazatricyclo[3,3,1,$1^{3,7}$]decane
Trade Name(s):	Aminoform, Formin, Uritone, Cystamin
Type of compound:	Formaldehyde donor, n-acetal

1. Structure and Chemical Properties

Appearance:	Crystals, granules. or powder; hygroscopic; volatile at low temperature
Odor:	Odorless
Solubility:	1 g dissolves in 1.5 ml of water or 12.5 ml of alcohol
Optimum pH:	pH of 0.2 M aqueous solution 8.4; forms formaldehyde at acid pH
Compatibility:	Compatible with anionics. cationics, nonionic detergents and proteins
Neutralization:	Dilution and peptone
Structural formula:	

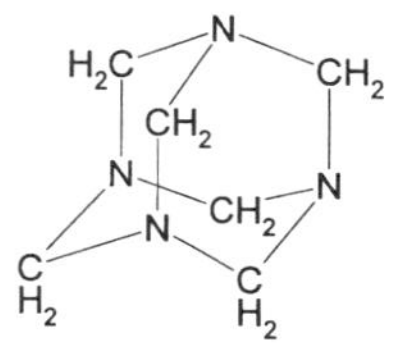

2. Antimicrobial Spectrum

Hexamethylenetetramine itself shows no antimicrobial activity but is derived from the breakdown into formaldehyde. See the data for formaldehyde.

3. Toxicity

Acute oral toxicity	LD_{50}, 200 mg/kg
Chronic toxicity:	90-day test (rat) with 0.4 g/day produced no toxic effect; 1% hexamethylenetetramine daily oral dose over 60 weeks showed no effect on mice and rats.
Carcinogenicity:	0.15 mg/kg per day without carcinogenic effect by oral application

4. Cosmetic and Other Applications

Preservation of lotions and creams (0.2%); if the concentration is higher than 0.05% free formaldehyde, warning "Contains Formaldehyde" must appear on the label. Preservation of hides; phenol-formaldehyde resin; corrosion inhibitor for steel.

5. Mode of Action

As formaldehyde is released from the breakdown of hexamethylenetetramine it denatures proteins by reacting with amino groups in proteins of the cell wall, membrane, and cytoplasm.

IMIDAZOLIDINYL UREA
EEC #: 36
CAS #.: 39236-46-9

Chemical Name(s):	N, N'-methylene-bis-N'-[1-(hydroxymethyl)-2,5-dioxo-4-imidazolidinyl] urea
Trade Name(s):	Germall 115. Blopure 100, Euxyl K 200
Type of compound:	Heterocyclic substituted urea; formaldehyde donor

1. Structure and Chemical Properties

Appearance:	Stable white powder
Odor:	Odorless
Solubility:	Water, 200; glycerol, 100; isopropanol, 0.05
Optimum pH:	Wide range
Stability:	Releases formaldehyde upon decomposition; decomposes >160°C
Compatibility:	Compatible with ionics. nonionics and protein
Neutralization:	Dilution and peptone
Structural formula:	

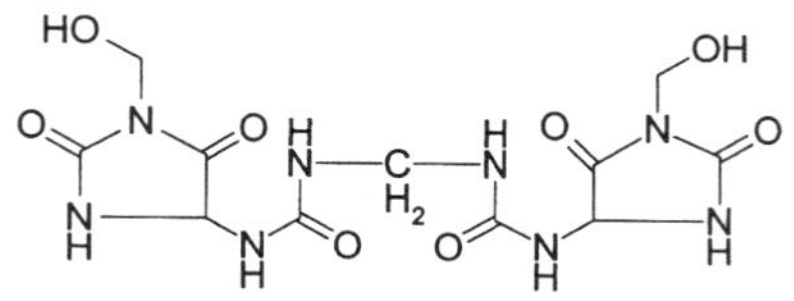

2. Antimicrobial Spectrum

Test organisms (10^6 CFU/ml)	Minimal inhibitory concentration (μg/ml) (serial dilution test; 24 to 72 hr)	Minimal germicidal concentration (μg/ml) (suspension test: 24 to 72 hr)
Staphylococcus aureus	1000	2000
Escherichia coli	2000	8000
Pseudomonas aeruginosa	2000	8000
Candida albicans	8000	8000
Aspergillus niger	8000	8000

3. Toxicity

Acute oral toxicity	Rat: LD_{50} 2.57 g/kg
Acute dermal toxicity	Rabbit: LD_{50} >2 g/kg
Primary skin irritation	Rabbit: 1 to 5% solution not a primary irritant
Rabbit eye irritation:	1-5% not an eye irritant

4. Cosmetic and Other Applications

Use concentrations, 0.1 to 0.5% in combination with parabens or other antifungal preservatives; used for lotions, creams, hair conditioners, shampoos, and deodorants.

5. Mode of Action

As the chemical degrades, it donates formaldehyde to the microorganism. The formaldehyde denatures proteins by reacting with amino groups in proteins of the cell wall, membrane, and cytoplasm.

CHLOROMETHYL ISOTHIAZOLIINONE AND METHYL ISOTHIAZOLINONE
EEC #: 45
CAS #.: 26172-55-4

Chemical Name(s):	5-chloro-2-methyl-4-isothiazoline-3-one and 2-methyl-4-isothiazoline-3-one
Trade Name(Supplier):	Kathon CG
Type of compound:	Two types of isothiazolinones and inorganic magnesium salts in water

1. Structure and Chemical Properties

Appearance:	Nonviscous liquid at 1.5%; light amber
Solubility:	Highly soluble in water, lower alcohols, and glycols
Optimum pH:	Optimum pH 4 to 8; stability reduced in systems of increasing alkaline pH (pH >8).
Stability:	Loses activity upon storage in some formulations at elevated temperatures
Compatibility:	Compatible with anionic, cationic, and nonionic surfactants and emulsifiers; inactivated by hypochlorite
Neutralization:	Use of dilution with protein-containing broths such as Letheen or peptone broth
Structural formula:	

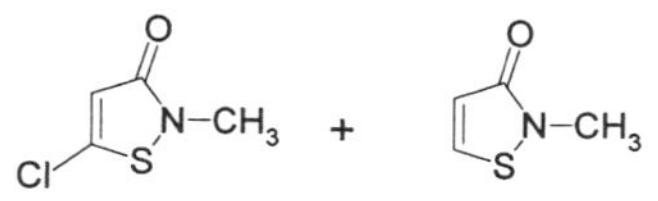

2. Antimicrobial Spectrum

Test organisms (10^6 CFU/ml)	Minimal inhibitory concentration (µg/ml) (serial dilution test; 24 to 72 hr)	Minimal germicidal concentration (µg/ml) (suspension test: 24 to 72 hr)
Staphylococcus aureus	150	500 to 1000
Escherichia coli	300	1500
Pseudomonas aeruginosa	300	1500 to 2000
Candida albicans	300	125
Aspergillus niger	200 to 600	—

Bactericidal activity: 5-chloro-2-methyl-4isothiazolon-3-one, 1.15%; Fungicidal activity: 2-methyl-4isothiazolin-3-one, 0.35%

3. Toxicity

Acute oral toxicity	Rat (male): LD_{50}, 3.35 g/kg
	Rat (female): LD_{50}, 2.63 g/kg
Acute dermal toxicity	Rabbit (male): LD_{50}, >5 g/kg
Subchronic toxicity	In a 3-month dietary study with rats, 2 g/kg per day nontoxic, no pathological findings; no effect level is 633 mg/kg-day. In a 3-month percutaneous 1 application/day, 5 days a week at 0.40%, Kathon was nontoxic; no pathological findings
Primary skin irritation	Rabbit: product diluted at 3.7% is a nonirritant
Human patch test:	0.37% Kathon CG (1OX minimum dose) shows no primary irritation or sensitization
Guinea pig sens.:	Skin sensitization (Magnusson-Kligman Test), 0.37%: Kathon CG showed no skin sensitization
Phototox/photosens.:	Neither phototoxic nor photosensitizing
Rabbit eye irritation:	18% corrosive to the eye, 3.7% no irritation
Teratology	Rat: No fetus toxicity; no teratogenicity in the range 100 to 1000 mg/kg
Mutagenicity	Ames Test, variable; cytogenetic study on male rats no effect at 19 to 190 and 1900 mg/kg

4. Cosmetic and Other Applications

Use concentrations, 0.035 to 0.15%; shampoos, hair conditioners. Hair and body gels. Color dye solutions, bubble baths, skin creams and lotions, mascaras. Used in cutting oils.

5. Mode of Action

Isothiazolinones inhibit active transport and glucose oxidation mainly by reacting with cellular thiol groups of proteins like ATPase and glyceraldehyde-3-phosphate dehydrogenase to denature them.

MONOMETHYLOL DIMETHYLHYDANTOIN (MDM HYDANTOIN) EEC #: 39 CAS #.: 28453-33-0

Chemical Name(s):	1-(Hydroxymethyl)-5,5-dimethylhydantoin
Trade Name(s):	Dantoin 685
Type of compound:	Formaldehyde donor

1. Structure and Chemical Properties

Appearance:	Crystals
Odor:	Odorless
Solubility:	Water soluble; soluble in ethanol or methanol
Optimum pH:	4.5 to 9.5
Stability:	Stable at temperatures below 85°C; formaldehyde is split off at pH 6 in aqueous solutions.
Compatibility:	Compatible with ionics. nonionics and protein
Neutralization:	Dilution and peptone
Structural formula:	

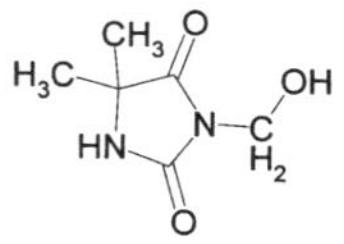

2. Antimicrobial Spectrum

Test organisms (10^6 CFU/ml)	Minimal inhibitory concentration (μg/ml) (serial dilution test; 24 to 72 hr)
Staphylococcus aureus	512
Escherichia coli	1024
Pseudomonas aeruginosa	2048
Candida albicans	2048
Aspergillus niger	2048

3. Toxicity
See data for DMDM Hydantoin

4. Cosmetic and Other Applications
Use concentration, 0.25%; EEC allows 0.2% for shampoos and as active compound in deodorants; biocide for cutting oils.

5. Mode of Action
See DMDM hydantoin.

DIMETHYL OXAZOLIDINE
EEC#: 51
CAS #.: 200-87-4

Chemical Name(s):	4.4- Dimethyl- 1,3 oxazolidine
Trade Name(Supplier):	Oxadine A
Type of compound:	Cyclic substituted amine, oxazolidine, formaldehyde donor

1. Structure and Chemical Properties

Appearance:	Colorless liquid
Odor:	Penetrating aminelike odor
Solubility:	Completely water soluble
Optimum pH:	6.0 to 11.0
Stability:	Unstable below pH 5.0
Compatibility:	Compatible with cationic, anionic, and nonionic systems over the pH range 5.5 to 11
Inactivation:	Low pH, dilution and peptone
Structural formula:	

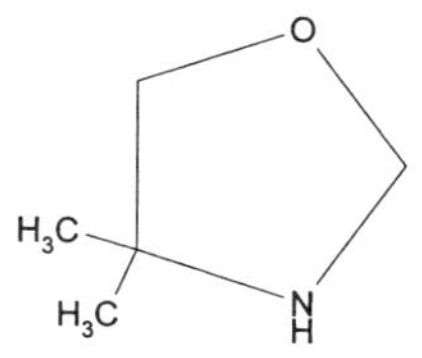

2. Antimicrobial Spectrum

Test organisms (10^6 CFU/ml)	Minimal inhibitory concentration (µg/ml) (serial dilution test; 24 to 72 hr)	Minimal germicidal concentration (µg/ml) (suspension test: 24 to 72 hr)
Staphylococcus aureus	125 to 500	500
Escherichia coli	250 to 500	500
Pseudomonas aeruginosa	250 to 500	500
Candida albicans	500 to 1000	1000
Aspergillus niger	250 to 1000	500

3. Toxicity

Acute oral toxicity	Male rat: LD_{50}, 950 mg/kg
Acute dermal toxicity	Rabbit: LD_{50}, 1400 mg/kg
Rabbit eye irritation:	5000 ppm in water has no discernible effect upon the eyes.
Human sensitization:	LC_{50}, 11.7 mg/liter; At use levels, no sensitization.
Mutagenicity:	Ames test, nonmutagenic

4. Cosmetic and Other Applications

Protein shampoos, hand creams: 0.05 to 0.2% use concentration; not in EEC guideline; antimicrobial preservative for cutting oils

5. Mode of Action

See DMDM hydantoin

PARABENS
EEC #: I/12
CAS #.:

Chemical Name(s):	p-Hydroxybenzolc acid ester
Trade Name(Supplier):	Solbrol; Nipagin; Nipasol
Type of compound:	Benzoic acid esters

1. Structure and Chemical Properties

Appearance:	White crystalline powder
Odor:	Odorless
Solubility:	In water 25°C (g/100 ml): 0.006, benzyl; 0.02, butyl; 0.04, propyl; 0.11, ethyl; 0.25, methyl
Optimum pH:	3 to 9.5
Stability:	Stable
Compatibility:	Compatible with anionic and cationic emulsions, proteins; incompatible with polyoxy-40-stearate, polyvinylpyrrolidone, methylcellulose
Neutralization:	Any nonionic emulsifier such as Tween 80
Structural formula:	R may be methyl, ethyl, propyl, butyl, or benzyl

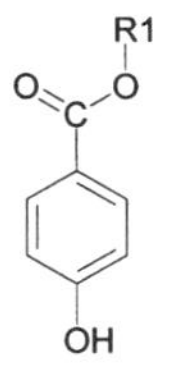

2. Antimicrobial Spectrum

Test organisms (10^6 CFU/ml)	Minimal inhibitory concentration (μg/ml) (suspension test; 24 to 72 hr)					Minimal germicidal concentration(μg/ml) (serial dilution test; 24 to 72 hr)				
	Methyl	Ethyl	Propyl	Butyl	Benzyl	Methyl	Ethyl	Propyl	Butyl	Benzyl
Staphylococcus aureus	800	500	150	120	120	1250	625	180	160	50
Escherichia coli	800	600	300	150	160	1250	1250	360	160	125
Pseudomonas aeruginosa	1000	800	400	175	160	1250	625	625	160	175
Candida albicans	1000	800	250	125	250	5000	2500	625	625	100
Aspergillus niger	600	400	200	150	1000	5000	5000	2500	1250	125

3. Toxicity

Acute oral toxicity	Dog:	Methyl LD_{50} = 3.0; ethyl LD_{50} = 5.0; propyl LD_{50} = 6.0; butyl LD_{50} = 6.0 (in g/kg)
Subchronic tox	Rabbit:	500 mg/kg-day methyl ester for 6 days without effect; 3000 mg with toxicity
Human toxicity:	2 g/day methyl and propyl ester over 1 month without effect	

4. Cosmetic and Other Applications

Use concentration 0.1% benzyl ester (EEC); 0.4% for a single ester and 0.8% for a mixture of esters; most commonly used: 0.18% methyl + 0.02% propyl ester. Preservative for pharmaceuticals and food.

5. Mode of Action

The most likely MOA supported by the data is disruption of membrane potential. This disruption interferes with membrane transport and energy generation. Cells exposed to parabens leak intracellular contents but show no overt change in cell structure and can recover when exposed to preservative-free media suggesting that cell lysis and membrane damage had not occurred.

POLYOXYMETHYLENE
EEC #: I/5
CAS #.: 9002-81-7

Chemical Name(s):	Polyoxymethylene
Trade Name(s):	Triformol, Formagene, Foromycen
Type of compound:	Poymerized formaldehyde

1. Structure and Chemical Properties

Appearance:	White Crystalline Powder
Odor:	Formaldehyde odor
Solubility:	Soluble in water. Insoluble in alcohol.
Optimum pH:	4 to 8
Stability:	Unstable in alkaline solutions. In hot water, formaldehyde is given off. Container must be kept tightly closed to prevent formaldehyde off-gassing.
Compatibility:	Compatible with anionic and nonionic detergents. Not compatible with proteins, ammonia, oxidants, or heavy metals.
Neutralization:	Dilution and peptone
Structural formula:	

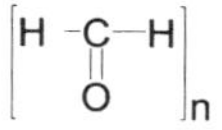

2. Antimicrobial Spectrum
Same as data for formaldehyde.

3. Toxicity
Same as data for formaldehyde.

4. Cosmetic and Other Applications
Used at a maximum of 0.2%. May be used as a disinfectant.

5. Mode of Action
Same as formaldehyde.

PHENETHYL ALCOHOL
MERCK INDEX #: 7094

Chemical Name(s):	2-phenylethanol
Trade Name(s):	
Type of compound:	Natural alcohol

1. Structure and Chemical Properties

Appearance:	Colorless liquid
Odor:	Rose or floral-like
Solubility:	2 ml in 100 ml water; miscible with alcohol (1:1 in 50% alcohol)
Optimum pH:	Acidic pH
Stability:	Instable in presence of oxidants
Compatibility:	Compatible with most surfactants; imparts a rose odor to formulae
Neutralization:	Partially inactivated with non-ionics and Tween 80
Structural formula:	

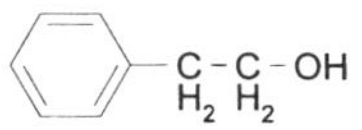

2. Antimicrobial Spectrum

Test organisms (10^6 CFU/ml)	Minimal inhibitory concentration (μg/ml) (serial dilution test; 24 to 72 hr)
Staphylococcus aureus	1250
Escherichia coli	2500
Pseudomonas aeruginosa	2500 to 5000
Candida albicans	2500
Aspergillus niger	5000

3. Toxicity

Acute oral toxicity	Rat:	LD_{50}, 1.79 g/kg
Acute dermal toxicity	Guinea pig:	LD_{50}, 5 to 10 ml/kg
Human eye irritation:	0.75%	
Guinea pig sens.:	No sensitization in concentrations of 1-2%	
Teratogenicity:	No effect	

4. Cosmetic and Other Applications

Eye makeup at 1%; pharmaceuticals: eyedrops; cutaneous antiseptic, 0.3% combined with 0.01% benzalkonium chloride; oral pharmaceuticals, 0.3% to 0.5%.

5. Mode of Action

Disruption of membrane by solubilization of lipids and possibly denatures proteins.

PHENOXYETHANOL
EEC #: 43
CAS #.: 122-99-6

Chemical Name(s):	2-phenoxyethanol
Trade Name(s):	Dowanol EPH; Phenyl Cellosolve; Phenoxethol; Phenonip
Type of compound:	Phenolic

1. Structure and Chemical Properties

Appearance:	Oily liquid
Odor:	Faint aromatic odor
Solubility:	2.67% in water; miscible with alcohol
Optimum pH:	Wide pH tolerance
Stability:	Highly stable
Compatibility:	Compatible with anionic and cationic detergents
Neutralization:	Inactivated by nonionics and dilution
Structural formula:	

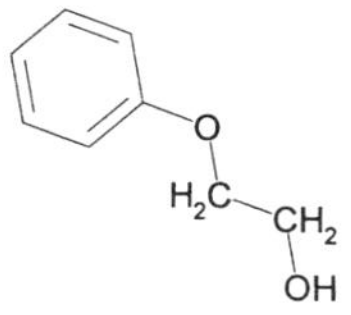

2. Antimicrobial Spectrum

Test organisms (10^6 CFU/ml)	Minimal inhibitory concentration (µg/ml) (serial dilution test; 24 to 72 hr)
Staphylococcus aureus	2000
Escherichia coli	4000
Pseudomonas aeruginosa	4000
Candida albicans	5000
Aspergillus niger	5000

3. Toxicity

Acute oral toxicity	Rat: LD_{50}, 1.3 g/kg
Primary skin irritation:	No irritation

4. Cosmetic and Other Applications

0.5 to 2.0% used in combination with parabens, dehydroacetic acid, or sorbic acid; may add with propylene glycol for improvement of water solubility. Used as a bactericide with quaternary ammonium compounds.

5. Mode of Action

Disruption of membrane by solubilization of lipids and possibly denatures proteins.

PHENYLMERCURIC ACETATE
EEC #: 11
CAS #.: 62-38-4

Chemical Name(s):	Phenylmercuric acetate
Trade Name(s):	Advacide PMA 18; Cosan PMA; Mergal A25; Metasol 30; Nuodex PMA 18
Type of compound:	Organic mercurial

1. Structure and Chemical Properties

Appearance:	Crystalline
Solubility:	Poorly soluble (1:600 in water); soluble in hot ethanol.
Optimum pH:	Neutral
Compatibility:	Incompatible with iodine compounds, sulfides, thioglycollates, anionics, halogens, ammonia. Compatible with nonionic emulsifiers.
Neutralization:	Use thiol medium and dilution
Structural formula:	

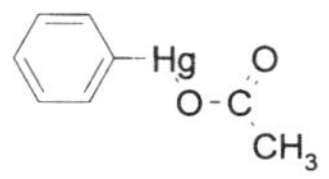

2. Antimicrobial Spectrum

Test organisms (10^6 CFU/ml)	Minimal inhibitory concentration (μg/ml) (serial dilution test; 24 to 72 hr)
Staphylococcus aureus	0.1
Escherichia coli	0.5
Pseudomonas aeruginosa	1 to 5
Candida albicans	8
Aspergillus niger	16

3. Toxicity

Acute oral toxicity	Rat: LD_{50}, 30 mg/kg (acetate salt); LD_{50}, 60 mg/kg (nitrate)
Chronic toxicity:	0.1 mg Hg/kg animal feed over 1 year shows no effect in rats; 0.5 mg Hg/kg shows kidney effects in some animals.
Primary skin irritation:	0.1% solution is a skin irritant; 0.01% shows no effect

4. Cosmetic and Other Applications

Limited to eye-area cosmetics that cannot be preserved with any other available preservative. Must be labeled, "Contains phenylmercurial compounds." Maximum concentration in EEC is 0.003% in creams and concentrated shampoos that cannot be preserved with any other option. Also used in pharmaceutical eyedrops and nasal sprays at 0.002 to 0.005%.

5. Mode of Action

Protein denaturant of cytoplasmic and membrane-associated proteins by heavy metal reaction with thiol groups of proteins.

O-PHENYLPHENOL
EEC #: I/7
CAS #.:90-43-7

Chemical Name(s):	ortho-phenylphenol
Some trade names:	Dowicide 1, Preventol O
Type of compound:	phenolic

1. Structure and Chemical Properties

Appearance:	White flaky crystals
Odor:	Mild characteristic odor
Solubility:	Virtually insoluble in water except the sodium salt (tetrahydrate) is soluble at 122 g/100 g water.
Optimum pH:	12 to 13.5 for the sodium salt.
Compatibility:	Incompatible with nonionics, carboxymethylcellulose, polyethylene glycols, quaternary compounds, and proteins.
Inactivation/Neutralization:	Dilution
Structural formula:	

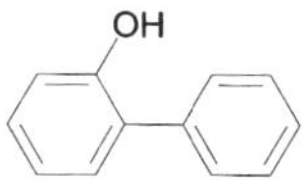

2. Antimicrobial Spectrum

Test organisms (10^6 CFU/ml)	Minimal inhibitory concentration (µg/ml) (serial dilution test; 24 to 72 hr)	Minimal germicidal concentration (µg/ml) (suspension test: 24 to 72 hr)
Staphylococcus aureus	100	625
Escherichia coli	500	1250
Pseudomonas aeruginosa	1000	1250
Candida albicans	50	125
Aspergillus niger	50	125

3. Toxicity

Acute oral toxicity	Rat:	LD_{50}, 2.48g/kg
Subchronic toxicity	Rat:	200 mg/kg-day over 32 days without toxic effects.
Chronic toxicity	Rat:	0.2% in feed over 2 years without toxic effects

4. Cosmetic and Other Applications

Used in a variety of products where "medicinal" odor is not a problem. Maximum use defined in EEC as 0.2%.

5. Mode of Action

As with many phenolics o-phenylphenol most likely uncouples oxidation from phosphorylation and inhibits active transport by disrupting the cell membrane through solubilizing lipids and denaturing proteins. Once membrane integrity is compromised, the cell is more permeable to protons and thus any potential gradient for running ATP synthetase is destroyed.

PROPIONIC ACID
EEC #: I/2
CAS #.: 79-09-4

Chemical Name(s):	Propionic acid
Trade Name(s):	Mycoban
Type of compound:	Organic acid

1. Structure and Chemical Properties

Appearance:	Oily liquid
Odor:	Pungent, slightly disagreeable racid odor.
Solubility:	Miscible with water; may come out of solution by addition of salts.
Optimum pH:	3.5 to 4.5; limit pH 6
Stability:	Stable
Neutralization:	Raise pH above pKa, dilution
Structural formula:	

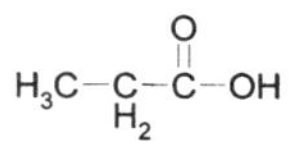

2. Antimicrobial Spectrum

Test organisms (10^6 CFU/ml)	Minimal inhibitory concentration (µg/ml) (serial dilution test; 24 to 72 hr)
Staphylococcus aureus	2000
Escherichia coli	2000
Pseudomonas aeruginosa	3000
Candida albicans	2000
Aspergillus niger	2000

3. Toxicity

Acute oral toxicity	Rat: LD_{50}, 2.6 g/kg
Subchronic toxicity:	No effect in rat feeding study over several weeks at 1 to 3% sodium or calcium propionate in feed.
Chronic toxicity:	3.75% addition to rat chow given to rats over 1 year without negative effects.
Primary skin irritation:	Irritates skin/mucous membranes when concentrated.

4. Cosmetic and Other Applications

2% allowed in EEC; usual use is in bread at 0.15 to 0.4% to prevent ropiness.

5. Mode of Action

Destroys chemiosmotic balance across the cytoplasmic membrane by disruption of the membrane electrical potential through dissociation of protons into the cytoplasm of the cell.

QUATERNIUM-15
EEC #: 48
CAS #.: 4080-31-3

Chemical Name(s):	cis isomer of 1-(3-chloroallyl)-3,5,7-triaza-1-azoniaadamantane-chloride, N-(3-chloroallyl)- hexammonium chloride
Trade Name(s):	Dowicil 200; Dowicide Q; Preventol D1
Type of compound:	Quaternary adamantane

1. Structure and Chemical Properties

Appearance:	Cream-colored powder; hygroscopic.
Odor:	Odorless
Solubility:	127 g/100 ml water; less soluble in propylene glycol (18.7%), glycerol (12.6%), ethanol (1.85%).
Optimum pH:	4 to 10
Stability:	Unstable above 60°C; yellows in some cream formulations.
Compatibility:	Compatible with anionics, nonionics, cationics, and proteins.
Neutralization:	Dilution and peptone
Structural formula:	

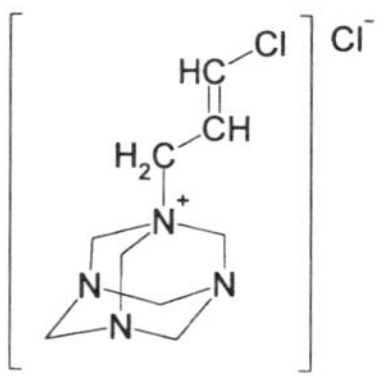

2. Antimicrobial Spectrum

Test organisms (10^6 CFU/ml)	Minimal inhibitory concentration (µg/ml) (serial dilution test; 24 to 72 hr)
Staphylococcus aureus	200
Escherichia coli	500
Pseudomonas aeruginosa	1000
Candida albicans	>3000
Aspergillus niger	>3000

3. Toxicity

Acute oral toxicity	Rat: LD_{50}, 0.94 to 1.5 g/kg
	Rabbit: LD_{50}, 40 to 80 g/kg
Primary skin irritation:	Not a primary skin irritant to human skin.
Guinea pig sens.:	Nonsensitizing at up to 2 %.
Mutagenicity:	Nonmutagenic based on Ames test and unscheduled DNA synthesis.

4. Cosmetic and Other Applications

In-use concentration, 0.02 to 0.3%. In EEC up to 0.2% in shampoos, and skin lotions. Preservative for paints.

5. Mode of Action

Donates formaldehyde without the release of gaseous formaldehyde. Formaldehyde alkylates amino and sulfhydryl groups of amino acids as well as the ring nitrogens of purine bases resulting in protein and DNA denaturation.

SALICYLIC ACID
EEC #: I/3
CAS #.: 69-72-7

Chemical Name(s):	2-hydroxybenzoic acid
Trade Name(s):	
Type of compound:	Organic acid

1. Structure and Chemical Properties

Appearance:	Crystalline powder; upon heating, it decomposes into phenol and CO_2
Odor:	Odorless
Solubility:	1 g in 460 ml water; increased solubility of the salts
Optimum pH:	2 to 5
Stability:	Discolors with iron salts; discolors in sunlight and must be protected from light exposure.
Compatibility:	Incompatible with iron salts and iodine.
Neutralization:	pH above pKa; dilution
Structural formula:	

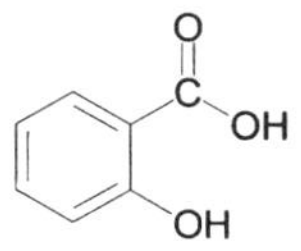

2. Antimicrobial Spectrum

Test organisms (10^6 CFU/ml)	Minimal inhibitory concentration (µg/ml)* (serial dilution test; 24 to 72 hr)
Staphylococcus aureus	1250
Escherichia coli	1250
Pseudomonas aeruginosa	2500
Candida albicans	2500
Aspergillus niger	2500

*Data are for pH of 3.2

3. Toxicity

Acute oral toxicity — Rat: LD_{50}, 891 mg/kg; Rabbit: LD_{50}, 1300 mg/kg

Toxicokinetic considerations: Salicylic acid is quickly resorbed but slowly metabolized and secreted running the risk of accumulation.

4. Cosmetic and Other Applications

Used at 0.025 to 0.2%. The EEC concentration limit is 0.5% and not to be used in products for children under 3. Also used as a preservative of food products in a few countries; it is forbidden for such use in others. Typically used as a topical keratolytic agent in anti-dandruff preparations.

5. Mode of Action

Destroys chemiosmotic balance across the cytoplasmic membrane by disruption of the membrane electrical potential through dissociation of protons from the compound into the cytoplasm of the cell. May also denature some enzymes.

SODIUM PYRITHIONE
EEC #: 40
CAS #.: 3811-73-2

Chemical Name(s):	Sodium-2-pyridinethiol-1-oxide
Trade Name(S):	Sodium Omadine, Pyrion-Na
Type of compound:	Cyclic thiohydroxamic acid salt; pyridine derivative

1. Structure and Chemical Properties

Appearance:	White to yellowish powder
Odor:	Mild odor
Solubility:	53 g/100 ml water; 19 g/100 ml ethanol.
Optimum pH:	7 to 10
Stability:	Unstable to light, oxidizing agents, and reducing agents.
Compatibility:	Inactivated by nonionics and chelated when reacted with metal ions.
Neutralization:	Tween 80 or thiol broth and dilution
Structural formula:	

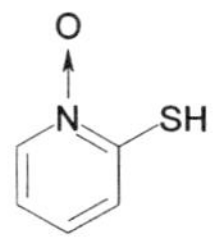

2. Antimicrobial Spectrum

Test organisms (10^6 CFU/ml)	Minimal inhibitory concentration (μg/ml) (serial dilution test; 24 to 72 hr)
Staphylococcus aureus	1
Escherichia coli	8
Pseudomonas aeruginosa	512
Candida albicans	4
Aspergillus niger	2

3. Toxicity

Acute oral toxicity	Rat:	LD_{50}, 875 mg/kg
	Mouse:	LD_{50}, 1172 mg/kg
Subchronic toxicity		
	Rat:	Oral administration of 75 mg/kg-day over 30 days or intraperitoneal administration of 40 mg/kg-day over 30 days produced no pathological effects.
Rabbit eye irritation:	No effect with 10% solution at pH 7.8 to 7.9	

4. Cosmetic and Other Applications

Use concentration 250 to 1000 ppm (as active); EEC guideline is 0.5% for rinse off products only. Used to preserve cutting oils at 0.005%

5. Mode of Action

Several studies have confirmed the hypothesis that pyrithiones destroy the chemiosmotic balance across the cytoplasmic membrane by disruption of the membrane electrical potential through dissociation of protons into the cytoplasm of the cell. However, there are likely to be other mechanisms extant that destroy the membrane integrity and thus membrane transport, in general. The presence of a reactive sulfhydril suggests it would denature membrane proteins for example.

S0RBIC ACID
EEC #: I/4
CAS #.: 110-44-1

Chemical Name(s):	2,4-Hexadienoic acid
Trade Name(s):	
Type of compound:	Organic acid

1. Structure and Chemical Properties

Appearance:	Crystalline
	Ca and K salts are white powder
Odor:	Faint characteristic odor
Solubility:	0.25% in water; 0.29% in ethanol; 8.4% in isopropanol; 0.5 to 1.0% in oils
	Potassium salt — 138 g/100 ml water
	Calcium salt — 1.2% in water
Optimum pH:	less than 6.5
Compatibility:	Incompatible with nonionics
Neutralization:	pH above the pKa and Tween 80
Structural formula:	

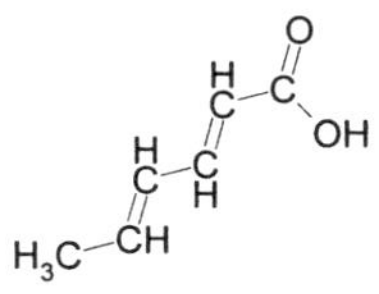

2. Antimicrobial Spectrum

Test organisms (10^6 CFU/ml)	Minimal inhibitory concentration (μg/ml) (serial dilution test; 24 to 72 hr)
Staphylococcus aureus	50 to 100
Escherichia coli	50 to 100
Pseudomonas aeruginosa	100 to 300
Candida albicans	25 to 50
Aspergillus niger	200 to 500

3. Toxicity

Acute oral toxicity	Rat: LD_{50}, 7.36 g/kg (acid) LD_{50},5.94 g/kg (sodium sorbate)
Chronic toxicity:	5% sorbic acid added to animal feed over a lifetime had no negative effects 10% over 2 years showed reduction in weight, but an increase in thyroid gland, kidney, and liver.

4. Cosmetic and Other Applications

Antifungal for creams and lotions. Preservative in pharmaceutical oral dosages, and in food and wine

5. Mode of Action

Destroys chemiosmotic balance across the cytoplasmic membrane by disruption of the membrane electrical potential through dissociation of protons from the compound into the cytoplasm of the cell.

TRICLOSAN
EEC #: 28
CAS #.: 3380-34-5

Chemical Name(s):	2,4,4′-Trichloro-2′-hydroxydiphenylether
Trade Name(s):	Irgasan DP 300, Triclosan
Type of compound:	Chlorinated diphenyl ether

1. Structure and Chemical Properties

Appearance:	White crystalline powder
Odor:	Faint aromatic odor
Solubility:	0.001% in water; 100% in propylene glycol and ethanol; poor solubility in alkali hydroxide
Optimum pH:	4 to 8
Stability:	Thermostable at 280°C; labile with bleaching agents or chlorine
Compatibility:	Widely compatible with most raw materials except non-ionics
Neutralization:	Tween and lecithin
Structural formula:	

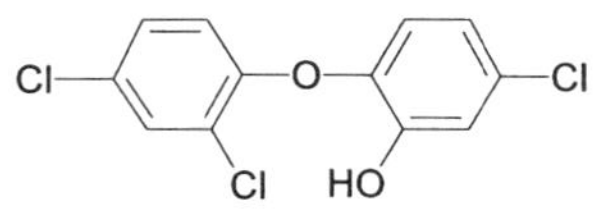

2. Antimicrobial Spectrum

Test organisms (10^6 CFU/ml)	Minimal inhibitory concentration (μg/ml) (serial dilution test; 24 to 72 hr)	Minimal germicidal concentration (μg/ml) (suspension test; 10 min)
Staphylococcus aureus	0.1	25
Escherichia coli	5	500
Pseudomonas aeruginosa	>300	—
Candida albicans	10	25
Aspergillus niger	100	—

3. Toxicity

Acute oral toxicity	Rat:	LD_{50}, 4.53 to >5 g/kg
	Dog:	LD_{50}, >5 g/kg
Subchronic toxicity:	Rat:	Oral administration of 0.45 g/kg-day over 4 weeks produced no toxic reactions.
Primary skin irritation:	Rat:	Cutaneous application over 4 weeks produced no irritation.
Rabbit eye irritation:	No irritation	
Photosensitization:	No effect	
Guinea pig sens.:	No sensitization	
Human sens.:	No sensitization	
Carcinogenicity:	Mouse:	Skin application 3 X/wk over 18 months produced no carcinogenic reaction

4. Cosmetic and Other Applications

In-use concentration of 0.1 to 0.3% used in deodorants, shampoos, and soaps as a bacteriostatic agent. EEC limit 0.3%. Also used as a skin disinfectant and preservative.

5. Mode of action

Like many chlorinated phenolics (e.g., tricarbanilide, pentachlorophenol, trichlorocarbanilide) Triclosan may uncouple oxidation from phosphorylation and inhibit active transport by denaturing membrane proteins. They do this by rendering the membrane permeable to protons and thus short-circuit the potential gradient or protonmotive force.

UNDECYLENIC ACID
EEC #: 13
CAS #.: 112-38-9

Chemical Name(s):	10-Undecenoic acid
Trade Name(s):	Declid, Renselin, Sevinon
Type of compound:	Organic acid; may also be used as the zinc salt or as a monoethanolamide-di-sodium- sulfosuccinate

1. Structure and Chemical Properties

Appearance:	Liquid or crystals
Odor:	Sweaty odor
Solubility:	Insoluble in water; soluble in alcohol
Optimum pH:	4.5 to 6.0
Compatibility:	Compatible with boric acid and salicylic acid
Neutralization:	pH above its pKa, dilution
Structural formula:	

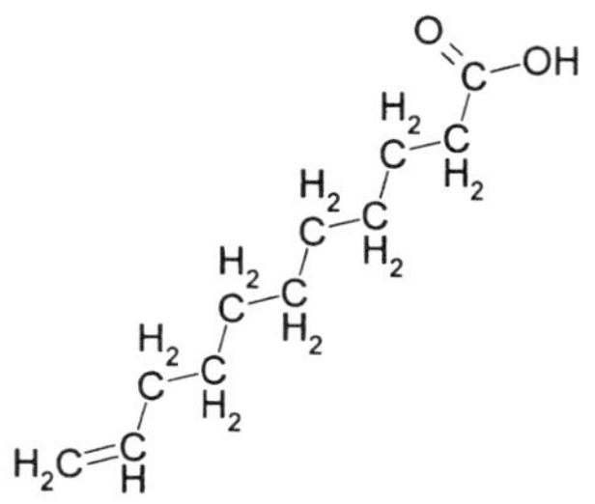

2. Antimicrobial Spectrum

Used as an antifungal agent at 2 to 15% in products for tinea pedis, capitis, and cruris, moniliasis, and vulvovaginitis. Antifungal nature is enhanced in the zinc salt form; activity greatest at acid pH.

3. Toxicity

Acute oral toxicity	Rat: LD_{50}, 2.5 g/kg
Primary skin irritation:	Irritant at >1% on mucous membranes

4. Cosmetic and Other Applications

EEC use limited to 0.2%. In the sulfosuccinate form (undecylenic acid monoethanolamide-di-sodium-sulfosuccinate), it is used at 1% in shampoos as an antidandruff agent at pH 5.0 to 6.5. Also used in pharmaceuticals as a topical antifungal at 2 to 15%.often in combination with boric acid and salicylic acid.

5. Mode of Action

Like many organic acids, it probably destroys chemiosmotic balance across the cytoplasmic membrane by disruption of the membrane electrical potential through dissociation of protons into the cytoplasm of the cell.

PIROCTONE OLAMINE
EEC #:57
CAS #.:68890-66-4

Chemical Name(s):	1-hydroxy-4-methyl-6-(2, 4, 4-trimethyl-pentyl)-2(1H) pyridone ethanolamine salt
Some Trade Name(s):	Octopirox
Type of compound:	Pyridone derivative

1. Structure and Chemical Properties

Appearance:	White to slightly yellowish powder
Odor:	Odorless
Solubility:	0.2% in water; 0.05 to 0.1% in various oils; 10% in alcohol; 1 to 10% in surfactants depending on solubilizing effects of surfactant.
Optimum pH:	5 to 9
Stability:	Stable up to 80°C for two weeks
Compatibility:	anionic, cationic, and amphoteric surfactants. Incompatibilities with some fragrances.
Neutralization:	dilution
Structural formula:	

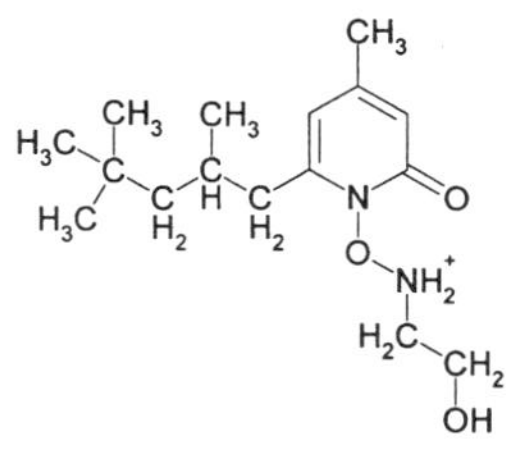

2. Antimicrobial Spectrum

Test organisms (10^6 CFU/ml)	Minimal inhibitory concentration (µg/ml) (serial dilution test; 24 to 72 hr)
Staphylococcus aureus	32
Escherichia coli	64
Pseudomonas aeruginosa	625 to 1250
Candida albicans	64
Aspergillus niger	—

3. Toxicity

Acute oral toxicity	Rat:	LD_{50}, 8.1 g/kg
	Mouse:	LD_{50}, 2.48g/kg
Acute dermal toxicity:	Rat and Dog:	NOEL 100 mg/kg-day
Primary skin irritation:	Low under use conditions	
Rabbit eye irritation:	Good tolerance	
Human sens.:	No sensitization induced	
Guinea pig sens.:	No sensitization induced	
Teratogenicity:	Not a teratogen	
Mutagenicity:	Not a mutagen (point and chromosomal mutation tests)	

4. Cosmetic and Other Applications

Antidandruff products, particularly transparent products. Use concentration is 0.5 to 1.0% in rinse off products, 0.05 to 0.1% in leave on products. Exhibits adsorbtion to the keratin and thus good substantivity.

5. Mode of Action

May disrupt membrane function by protein denaturation. There may be other mechanisms extant that destroy the membrane integrity and thus membrane transport, in general.

ZINC PYRITHIONE
EEC #: I/8
CAS #.: 13463-41-7

Chemical Name(s):	Zinc bis-(2-pyridinethiol-1-oxide)bis-(2-pyridylthio)zinc-1,1'-dioxide
Trade Name(s):	Zinc Omadine, Vancide
Type of compound:	Cyclic thiohydroxamic acid

1. Structure and Chemical Properties

Appearance:	White to yellowish crystalline powder
Odor:	Mild odor
Solubility:	Soluble in water, 15 ppm; in ethanol, 100 ppm; in PEG 400, 2000 ppm.
Optimum pH:	4.5 to 9.5
Stability:	Forms an insoluble product with some cationics and amphoterics; unstable to light and oxidizing agents; unstable in acid or alkaline solutions at higher temperatures
Compatibility:	Incompatible with EDTA which chelates out the zinc; slight inactivation by nonionics
Neutralization:	Tween 80 and use of dilution in thiol broth
Structural formula:	

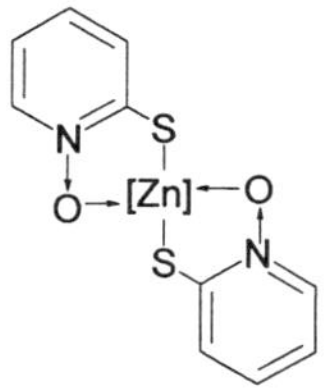

2. Antimicrobial Spectrum

Test organisms (10^6 CFU/ml)	Minimal inhibitory concentration (μg/ml) (serial dilution test; 24 to 72 hr)
Staphylococcus aureus	4
Escherichia coli	16
Pseudomonas aeruginosa	512
Candida albicans	0.25
Aspergillus niger	2

3. Toxicity

Acute oral toxicity	Rat:	LD_{50}, 200 mg/kg
	Mouse:	LD_{50}, 300 mg/kg
Subchronic toxicity	Rat:	10 ppm as food additive over 30 days showed no effect; higher doses toxic.
Chronic toxicity:	No irritation to humans or rabbits applied to skin in ointment form ion	
Primary skin irritation:	48% dispersion and powder are irritating to the skin and extrememly irritating to eyes.	
Guinea pig sens.:	Not an allergic sensitizer	

4. Cosmetic and Other Applications

Use concentration is 250 to 1000 ppm in various dandruff products; EEC guideline is 0.5% for rinse-off products only. Used in sanitization of paper goods, woolens, hospital goods

5. Mode of Action

Several studies have confirmed the hypothesis that pyrithiones destroy the chemiosmotic balance across the cytoplasmic membrane by disruption of the membrane electrical potential through dissociation of protons into the cytoplasm of the cell. However, there are likely to be other mechanisms extant that destroy the membrane integrity and thus membrane transport, in general. The presence of a reactive sulfhydril suggests it would denature membrane proteins for example.

INDEX

D

E

F

Q

T

U

V

W